DESIGN BRIEFS

TEACHERS' RESOURCE BOOK WITH PHOTOCOPY MASTERS

ROY LEE
Horringer Court Middle School, Bury St Edmunds, Suffolk

JOHN ALDRIDGE
Horringer Court Middle School, Bury St Edmunds, Suffolk

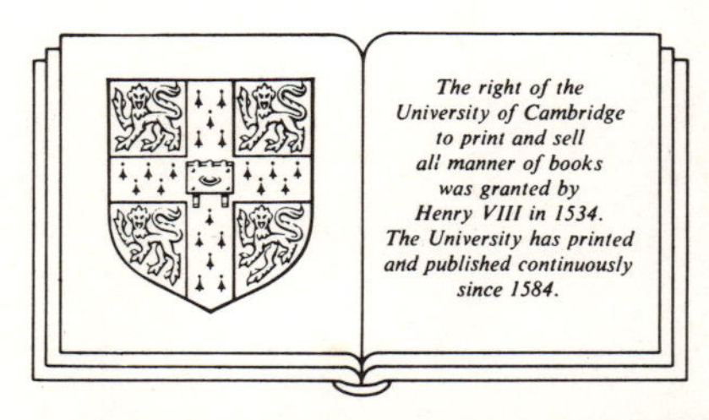

CAMBRIDGE UNIVERSITY PRESS
Cambridge
New York Port Chester Melbourne Sydney

For Carol and Marian

PHOTOCOPYING NOTICE

Published by the Press Syndicate of the University of Cambridge
The Pitt Building, Trumpington Street, Cambridge CB2 1RP
40 West 20th Street, New York, NY 10011, USA
10 Stamford Road, Oakleigh, Melbourne 3166, Australia

First published 1989

Printed in Great Britain by David Green Printers Ltd,
Kettering, Northants.

British Library cataloguing in publication data

Lee, Roy
Design briefs.
Teachers' resource book with photocopy masters.
1. Engineering. Design
I. Title II. Aldridge, John
620.'00425

ISBN 0 521 34826 9

VN

CONTENTS

INTRODUCTION

In recent years the objectives and assumptions of the traditional craft course have been increasingly called into question. New technologies in industry have led to a significant decline in demand for the traditional skills. This de-skilling process has even affected the amateur with, for example, the development of new adhesives and the availability of inexpensive multipurpose jigs for the DIY market. In the skill-orientated course it was assumed that the process of skill acquisition began at the start of the student's secondary career and continued beyond it. In fact, many students failed to complete this process and tended to acquire a narrow range of specific skills with only a limited ability to apply them.

The de-skilling process has been accompanied by a dramatic change in employment patterns. As the rate of technological change continues to increase, so our young people can expect to re-train several times during their working lives. Thus the 'skill for a lifetime' traditional apprenticeship system is changing, and so too are the assumptions underlying the craft course. A wide range of new skills have now become prerequisites of personal and national success. Not least amongst these is the ability to respond flexibly, imaginatively and effectively to new situations.

The post-war performance of Great Britain Ltd. in terms of innovation has been excellent. A recent Japanese survey estimated that 50 of the 100 most successful industrial innovations have been British. However, our record in the successful commercial exploitations of these ideas has been abysmal. We have failed to produce the engineers and technologists capable of converting innovative ideas into marketable products. Amongst the reasons for this failure is the low status accorded to engineers in Britain. Another factor has been the prescriptive nature of the traditionally taught craft subjects which tended to alienate the more thoughtful students.

Thus, industrial de-skilling and changes in employment patterns have been accompanied by a growing demand for a new generation of engineers capable of identifying needs and of developing products to meet those needs. The design–make–test–modify–evaluate model of the new Craft, Design and Technology courses is a direct response to that demand. This practical problem-solving approach is central to the philosophy of the new GCSE courses in CDT.

Design Briefs have been developed to provide a practical foundation course for both GCSE and CPVE courses. They have been designed to capture the imagination of boys and girls of all abilities in the 10–14 age range. Most of the Briefs have been successfully used in the middle school and some sections are suitable for use with top primary children. A guide to the suitability of each section for certain age groups is given in the next section. The Briefs set design problems which challenge the student to explore a wide range of options in seeking solutions. A databank of Information Sheets is made available through the Research Book. The entire package is intended to promote effective exploration throughout all stages of the design process.

The fertile imagination of children often exceeds their ability to realise their ideas in practical terms. They are hampered by low skill levels and by the inability to recognise their own weaknesses. As a result, it is easy for them to become discouraged by poor results in realisation. As discussed earlier, the traditional craft course overcame this problem by focusing on manual skills and confining the students to a series of skill-building exercises. By contrast, *Design Briefs* limit the tool skills required of the student whilst emphasising the development of problem-solving skills. The Briefs provide a structured series of problem-solving experiences which build confidence and ability in the wide range of design skills demanded by the GCSE course. The student works through a range of Briefs which, though closed and directed, are presented in a format which is fresh to the child and which both challenges and stimulates investigation. A wide range of knowledge, skills and problem-solving strategies is developed, thus equipping the student to tackle the more open-ended problems of the GCSE course. Some of the Design Briefs and a number of the Information Sheets also provide opportunities for able students to extend into open-ended problem-solving. The Information Sheets also form the basis of investigation into specific principles.

The philosophy underlying the *Design Briefs* package is that of CDT being essentially a practical problem-solving process. Thus the Briefs focus on a practical challenge which poses a series of questions for which immediate answers of both a practical and intellectual nature are required. In this 'now' situation the student is highly motivated to seek answers. Whilst the student should, of course, be encouraged to search as widely as possible for answers, the students' Research Book provides an immediately accessible databank of information.

The carefully structured format of the Briefs is designed to ensure success both in design and realisation. The confidence so generated will promote the further development of the student towards autonomy in the problem-solving process. The ultimate objective of the course is that all students, whether they are to proceed to the GCSE course or undertake more vocational work, will experience every aspect of the design process and through this experience be more discriminating in their appreciation both of their own efforts and of the work of others. It does not matter whether the focus of the students' attention is a finely cut dovetail joint or an aerodynamically efficient car body shell, a dust-proof plastic computer housing or an effective example of packaging graphics; the students' perception of their world and its artefacts will have been permanently enhanced and their appreciation enriched.

THE COURSE MATERIALS

THE STUDENTS' RESEARCH BOOK

The databank of 62 Information Sheets is arranged in six sections, as summarised below. A detailed list of the sheets appears in the Contents list.

Section	Age guide	Code	Total sheets in section
Structures	10–12	STR	8
Mechanisms	10–12	MEC	12
Using Science	10–14	SCI	6
Energy & Control	11–13	E&C	7
Electronics	12–14	EL	18
Using Materials	11–14	MAT	11

Readability and technical accuracy

In the preparation of the Research Book, a prime objective has been to ensure that the material is accessible to students throughout the ability range. It is intended that the average child of 10 will be able to read the entire contents of the Research Book. Careful attention has been paid to layout, sentence construction and overall reading levels. The aim has been to enable the student independently to research information for a Brief. There has from time to time been a conflict between strict technical accuracy and readability. Where this has occurred, we have deemed it desirable sometimes to simplify ideas at the cost of strict scientific exactitude.

Age guide

The age guide does not refer to the reading levels but to the suitability of the content of the Information Sheets and their associated Design Briefs. An important point to bear in mind is that, while few younger children would cope with the Electronics section, many older students can benefit from tackling earlier sections. For example, the Structures section is particularly useful for establishing the design–make–test–evaluate–modify model with older children who have had little previous experience of the CDT process.

Additional projects

A number of the Information Sheets make specific suggestions for practical investigation. In addition, it is also anticipated that teachers and students will generate their own projects. Design Briefs are not intended as a prescriptive end in themselves. Indeed, one measure of their success will be the amount of secondary briefs that they and the Information Sheets generate.

The Electronics sheets

The Electronics section is, in many ways, the most sophisticated of the course. A major feature of this section is that it is structured sequentially. In this respect it is unlike any other section. It is therefore most inadvisable for any student to study random pages of the Information Sheets. It is also vital that the student has pre-experience of the appropriate Information Sheets in this section, including practical investigations, before tackling any Electronics Design Brief.

It will be noticed that there are only four Electronics Design Briefs but that there are 18 Information Sheets. This is because many of the Information

Sheets are virtually Design Briefs in themselves. Indeed, it may be deemed desirable for some students to carry out all the practical investigations of the Information Sheets without attempting any of the Design Briefs. Most students will bring much less knowledge and experience to this section than to the others. Development towards independence in the student's approach to the Electronics Design Briefs can only be achieved by sequential and practical experience of the Information Sheet tasks.

The photographs

A wide variety of photographs from various sources has been included in the Research Book. Their purpose is to encourage students constantly to relate their own work to that of the broader world of design. The photographs include a number of examples of the practical application of technical and scientific principles. It is useful for students to create an exhibition of their own examples and to include sketches and photographs of examples in their design folder.

The glossary

Apart from acting as an initial reference source for students, the glossary is also useful as a tool for revision of key concepts. For the teacher it is a useful checklist of concepts covered by the course.

The index

The index provides a reference to the section and sheet number where the topic is discussed.

THE TEACHERS' RESOURCE BOOK

Resources supplements 1–6

The Resources Supplements are intended only to offer guidelines. It is expected that teachers will wish to adapt them to suit their own requirements and resources. The CDT philosophy of experimentation, evaluation and modification applies to teachers as well as to students! From time to time teachers new to CDT may experience problems in the construction of test equipment and, indeed, with the Briefs themselves. Should this be the case, colleagues and parents can prove to be a major resource. At the same time, the skills of the students themselves should not be overlooked.

The Electronics Circuit Board

Whilst teachers and students may well wish to design their own circuit board, this model is cost-effective and is suitable for all the electronics work of the *Design Briefs* course.

CDT Flow Diagram

The diagram is intended for distribution early in the course. However, it is useful to refer students back to the diagram from time to time in order to reinforce the philosophy of the course.

Teachers' Project Planner

The Planner has been prepared with non-specialists in mind, but more-experienced CDT teachers have also found it useful as a checklist.

Instruction sheets

The sheets are intended to provide additional information for students undertaking projects in associated Design Briefs.

Students' support materials

Students' Project Evaluation Guide and Sheet

These have been designed to offer students guidance in evaluation skills without resorting to the 'fill in this box' approach. Teachers may well find that a simpler version of the guide is appropriate for younger students, but the sheets have been successfully used with 11 and 12 year olds.

Student's Design Folder Guide

The Design Folder is a vital part of the course. In addition to recording the progress of the Brief, students are given an opportunity to reflect on their experience and so to reinforce it. The folder also provides a useful means for students to record their observations of design within their own environment. It is important for students to discuss the contents of their folders with their peers as well as with their teacher.

Photocopying

Before photocopying, please read the Photocopying Notice on page 2.

DESIGN BRIEFS IN PRACTICE

DESIGN BRIEFS AND THE NON-SPECIALIST TEACHER

Provided that non-specialist teachers have some practical experience of basic tool skills, there is no reason why they should experience any great difficulty in using the Design Briefs. The structured, child-orientated format of the Briefs, together with the information bank of the Research Book, provide most of the guidance required by the student. Indeed, because of the cross-curricular nature of the Briefs, teachers from other disciplines will often bring with them strengths particularly appropriate to the course. It is however desirable that, where possible, teachers try out each Design Brief. Insights gained here, especially by the non-specialist, can be particularly useful in judging when to intervene during the students' investigations.

INTRODUCING THE BRIEFS

In the first instance, it is recommended that only one Brief at a time is introduced to the class. This is particularly important when either the teacher or the class is unfamiliar with the Design Brief method. Whilst this limits the choice for students, it also gives them the opportunity of sharing their investigations and of comparing their solutions. For the same reason, if at a later stage two or three Briefs are undertaken by a class at the same time, it is desirable that only Briefs from the same section be chosen.

TEACHER INTERVENTION

An important reason for initially introducing only one Brief at a time is in order to reduce the logistical problems associated with the availability of materials, parts and tools. During practical sessions it is vital that the teacher is free to monitor the progress of individual students. As each student tackles the Brief, constant judgements need to be made concerning the moment to intervene. To do so too early will inhibit the development of the student's problem-solving skills; students must be given the time to find new solutions to old problems. However, it is vital to intervene in order to avoid the negative effects of frustration and failure.

Intervention by the teacher will take many forms. Sometimes a few words of encouragement will suffice, whilst on other occasions it will be necessary to refer a student back to a pertinent Information Sheet. Sometimes a major principle will require re-explanation by the teacher, at others it will be necessary to direct the student's investigations in a more rewarding direction. Often, a student's problem can be dealt with by a conversation with a peer. Whatever form the intervention takes, one of the principal roles of the teacher is the overt and covert management of each student's progress.

PRACTICAL POINTS

Materials

The thematic structure of the *Design Briefs* course lends itself to the planned purchasing of materials. When introducing the course it is unnecessary immediately to purchase a wide range of expensive materials. As each section of Briefs is introduced, so materials can be systematically acquired and storage systems established.

Teacher-prepared materials bank

Where Briefs require the preparation of materials by the teacher, it is time-effective to build up a bank of such items. Such a bank enables students who complete projects quickly to undertake supplementary projects independently.

Storage of students' work

It is obviously important that students can safely store and quickly retrieve their own work at the start and end of sessions. Many of the projects lend themselves to storage in plastic stacking bins. Cheap versions of these bins, available from DIY centres, have been found adequate for this purpose.

Design Brief sheets and other photocopied materials

There is obviously a considerable amount of photocopying involved during the introduction of the *Design Briefs* course. However, if sheets are collected at the end of each session and stored in document wallets, a library of such materials can be speedily established.

Equipment

Little sophisticated equipment is required for the *Design Briefs* course. A list of basic stock is, however, set out below.

Saws ○ Rasps ○ Files ○ Chisels
Wire strippers ○ Screwdrivers (small)
Steel rulers ○ Pliers ○ Set-squares
Disc sander
Hegner jigsaw
Strip heater (for plastics)
Belling oven (for plastics)
Pillar drill
Soldering iron
Rotatrim

THE DESIGN PROCESS

DURATION OF PROJECTS

Assuming that $1\frac{1}{2}$–2 hours are available for CDT each week, most Design Briefs will normally take half a term to complete. A few Briefs will only take 3–4 weeks and the Structures Briefs will require only one or two lessons, unless they are to be used as a major introduction to the design process itself.

INTRODUCING THE THEME

It is important that students gain an overview of the theme to be investigated at the start of each project. In this process artefacts, models and photographs are most useful. The establishment of a departmental museum of artefacts and mechanisms can be most effective as a focus for environmental investigation. Opportunities to disassemble and examine mechanisms will sometimes be appropriate.

INTRODUCTION OF THE BRIEF

Copies of the Brief are distributed and read through, either by the teacher or silently by the students. Any terms or concepts that the students do not understand are discussed.

PRELIMINARY INVESTIGATION

Students decide upon the areas within which they require further information. They are guided to the appropriate section of the Research Book by the index system. The teacher decides how much help, if any, is required by each student in this process.

Investigation can later be widened to other school-based resources or research at home. Wider research can also be undertaken, for example, by contacting commercial companies or through environmental investigation. In looking at products within the environment it is helpful to encourage students to pay particular attention to graphics, packaging and presentation. Sketches, photographs and notes in their design notebooks will all play a part as students record their observations and discoveries. At the conclusion of these investigations, discussion in groups is most effective in highlighting and reinforcing major observations.

BRAINSTORMING

The widest possible variety of potential responses to the Design Brief should be examined. All suggestions are valid at this stage and no idea is too far-fetched. It is important that a free-ranging and fast-moving discussion takes place either as a class or in small groups. At this point we are not interested in detail, only in broad outlines.

CONSIDERATION OF CONSTRAINTS

Initially in groups, and later on as a class, the constraints of time, materials, skill, money, safety and convenience need to be carefully considered. The pace is now slower and the discussion more considered. Now it is necessary to justify a point of view. The cognitive, linguistic and social skills being developed at this stage are very different to those used during brainstorming.

POTENTIAL SOLUTIONS

A range of possible solutions is now drawn up. Useful planning aids here will include squared paper, tracing paper, templates, and sometimes drawing instruments.

THE CHOSEN SOLUTION

After much discussion and careful analysis of the relative strengths and weaknesses of potential solutions, each student makes a final selection. At this point it is often desirable to produce a mock-up. It is important to consider such points as the thickness of materials when testing mock-ups. Prototyping using cheaper substitute materials, such as corrugated card, to test out structural aspects can save a great deal of time and money.

REALISATION

Realisation can be undertaken by individuals, pairs or groups. Much social and language development is possible where students work in groups. Materials are distributed and students progress at their own pace. At each stage, early finishers are guided to the next stage and they in their turn can help to guide others. Much consolidation of learning takes place as older students adopt the role of teacher. From time to time, especially at the start and end of lessons, a comparison of progress and problems encountered is desirable. To avoid becoming locked-in to failure, students need to test and evaluate the progress of their design at regular intervals. Modifications are then possible and an upwards spiral of success is assured. The timing of intervention by the teacher is discussed in the section Design Briefs in Practice.

EVALUATION

Evaluation can be undertaken as an individual, a group or class. One approach is an introduction to the evaluation process as a class, followed by detailed discussion in groups. Finally, students work as individuals in the completion of an evaluation sheet.

REFERENCE GUIDE

Design Brief title	No.	Key area	Content
Paper Structures	BST 1	Structures	paper bridges
Paper Structures	BST 2		paper pillar
Paper Structures	BST 3		paper tower
Strawmobile	BST 4		card vehicle (dynamic forces)
Eggtor Protector	BST 5		dynamic forces
Loopy Links	BME 1	Mechanisms	linkages
Make a Scene	BME 2		linkages, sequence
Push–pull Toy	BME 3		linkages, cranks, cams
Pinball Crazy	BME 4		linkages, microswitches
Catch in the Box	BME 5		box construction, levers, linkages
Marblous Boats	BSC 1	Using Science	buoyancy
Topping Ideas	BSC 2		balance, momentum, graphics, testing, evaluation
A4 Flying Wing	BSC 3		simple principles of flight
Simpli-kite	BSC 4		simple principles of flight
Go Fly a Kite	BSC 5		simple principles of flight, graphics
Weighing It Up	BSC 6		weighing device
Hot Rubber	BEC 1	Energy & Control	rubber-powered card vehicle
Hard-ships	BEC 2		rubber-powered ship
Ringcan Special	BEC 3		power and transmission
Ringcan Special (Plan Sheet)	BEC 3a		—
Ringcan Special II	BEC 4		power and transmission, gears, steering
Fairground Fun	BEC 5		fairground machines and mechanisms
Steady Freddy	BEL 1	Electronics	simple circuit with LED
Sensored	BEL 2		sensor device
About Time	BEL 3		time-delay circuit
Tweet Defeat	BEL 4		mechanical/electronic bird scarer
Mobile Magic	BGP 1	Graphics & Packaging	balance, moments and graphic design
Fun and Games	BGP 2		graphic design in card
Noughts & Crosses	BGP 3		graphic design in wood and acrylic
Noughts & Crosses (Plan Sheet)	BGP 3*a*		—
Pocket Games	BGP 4		graphic design in acrylic
Rock 'n' Roll	BGP 5		balance and graphic design

Age	Research Book reference						
	Structures	Mechanics	Using science	Energy and control	Electronics	Using materials	Information sheets
10–12	1–3						
	1–3						
	1–3						
	5, 7						
	1–3, 5, 6						
10–12		1–3					1
		3, 4			3, 4, 9	8–10	1
		3, 4, 7–9				8–10	1
		3, 4, 7			3, 4, 9	8–10	
		2, 4, 10				4, 5, 8–10	4
10–14			3				
			1, 4				
			2				
			2				
		1, 6, 7	2				
11–13	5, 7			3			
			2–4	3			2
	7			1–4			3
	7	7, 8, 11, 12		1–4			
		7–10		1,3,4			
12–14				1–9			
				1–12			
				1–15			
		3, 4	6	1–17			
12–14	6–8		1			4–9	
						4–10	4
						4–6	
			6				
			1, 2				

RESOURCES SUPPLEMENT 1

PAPER STRUCTURES

Test bed

Two or three of these stations are usually sufficient for a practical group of about 20. Dimensions and materials may be changed to suit the available resources.

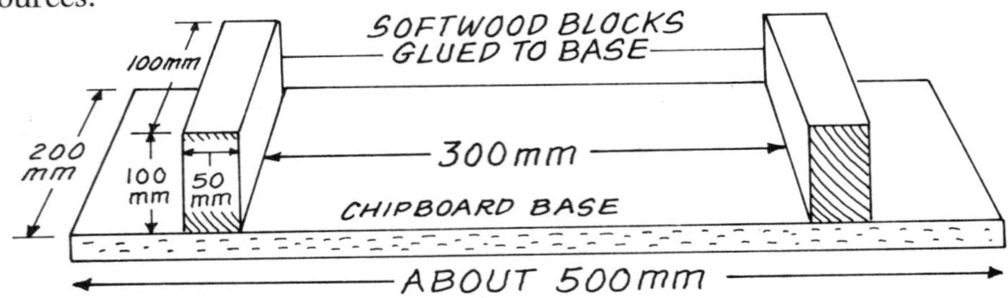

STRAWMOBILE

The ramp supplied need not be very sophisticated. The main requirement is that there should be as small a step as possible where the ramp meets the floor.

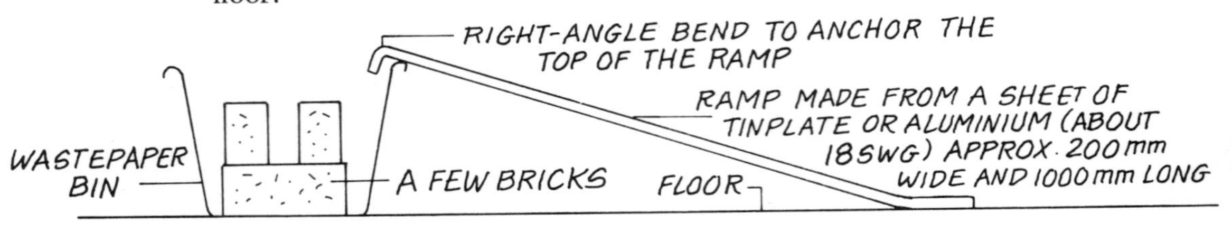

It is helpful to mark intervals of 1 m along the floor with coloured adhesive tape, starting from the bottom end of the ramp. This speeds measurement of performance considerably!

HARD-SHIPS

Test tank

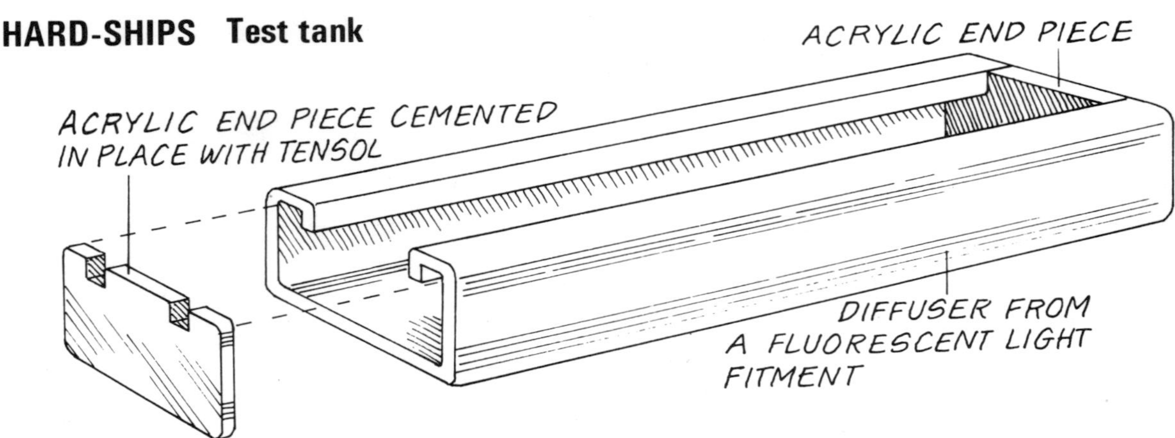

The tank used to test the designs needs to hold water to a depth of about 100 mm. It should be about 150 mm wide and about 1200–1800 mm long. A suitable tank can be made as shown from a fluorescent light diffuser with the ends blocked by acrylic 6 mm thick cut to shape with a power jigsaw and cemented in place with Tensol. Alternatively it is possible to use marine ply about 10 mm thick, screwed together, painted and sealed with an appropriate sealant (bathroom or shower type).

RESOURCES SUPPLEMENT 2

RINGCAN SPECIAL

Metal template

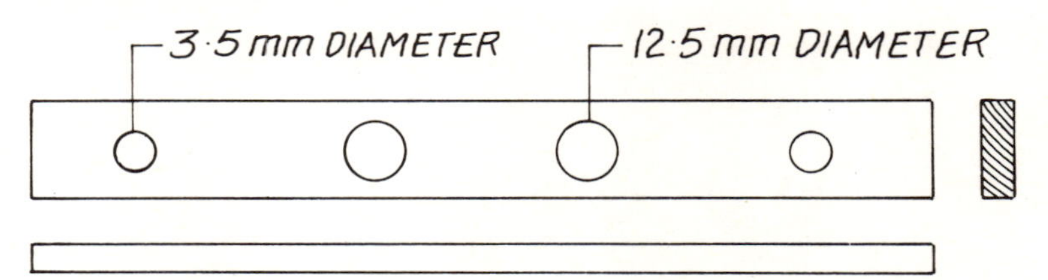

Use a strip of mild steel 20 mm wide by 5 mm thick by 200 mm long. Drill the holes as shown on the plan. If students use 2.5 or 3 mm drills for the axle holes and a 12 mm (or 15/32 inch) drill for the dowels then the clearance allowed should be adequate.

Suggested materials (per student)

Softwood, planed, 450 mm off 20 × 15 mm section (finished size).
Hardwood dowel $\frac{1}{2}$ inch (12 mm) – lengths depend on can size.
Welding rod 2.4 mm diameter, length 30 mm more than dowel length.
Plastic sleeving (PVC) 2 mm bore readily available from electronic component suppliers.
Tubing for spacers, clear PVC tubing as used in science laboratories or from aquaria suppliers. (Bore is not critical but 6.5 mm is suggested.)

NOTE: for some types of drive it will be necessary to have the can and the axle turn as one. To do this, remove the can and axle from the chassis, put a generous blob of general-purpose hot glue on to the axle near to the end of the can with the cardboard disc. Quickly slide the PVC spacer over the glue and up to the card disc and allow to set. It is now possible, for instance, to push a dressmaking pin through the tube; cut the pin to a suitable length so that a string with a looped end may be attached (see Energy and Control 1 and 3). Alternatively, pulleys or gears with integral bushes and grub screws (or push-fit) may be fitted to the axle.

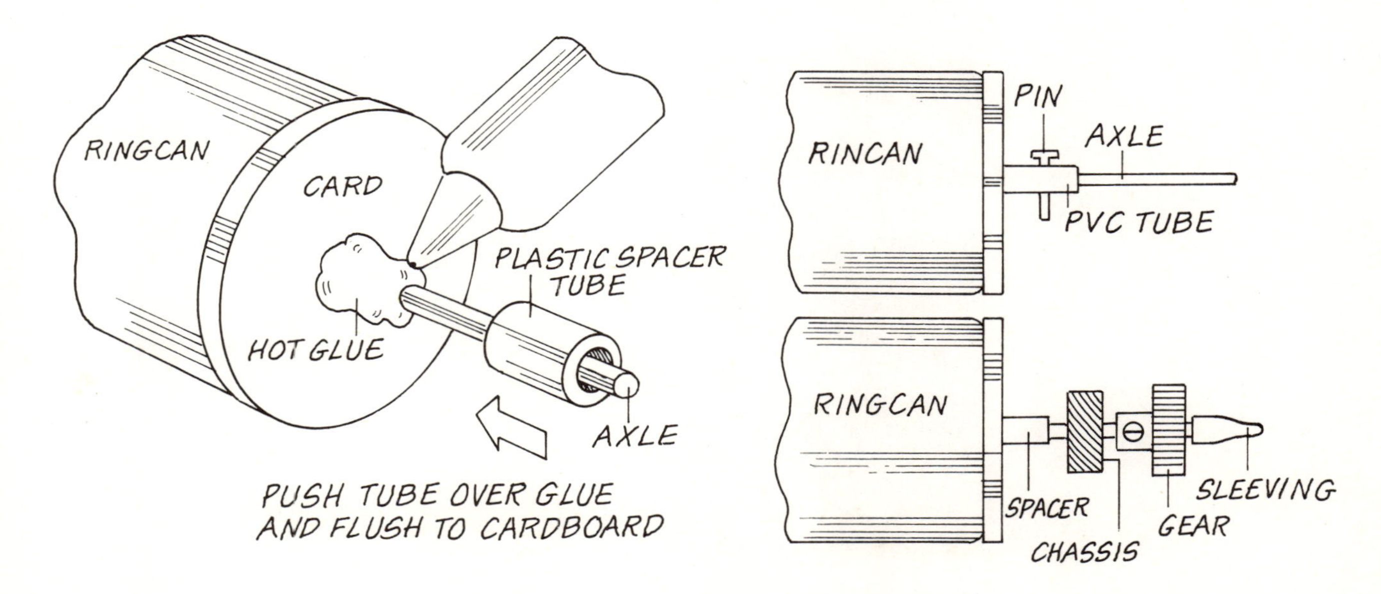

RESOURCES SUPPLEMENT 3

RINGCAN SPECIAL II The adjustable ramp

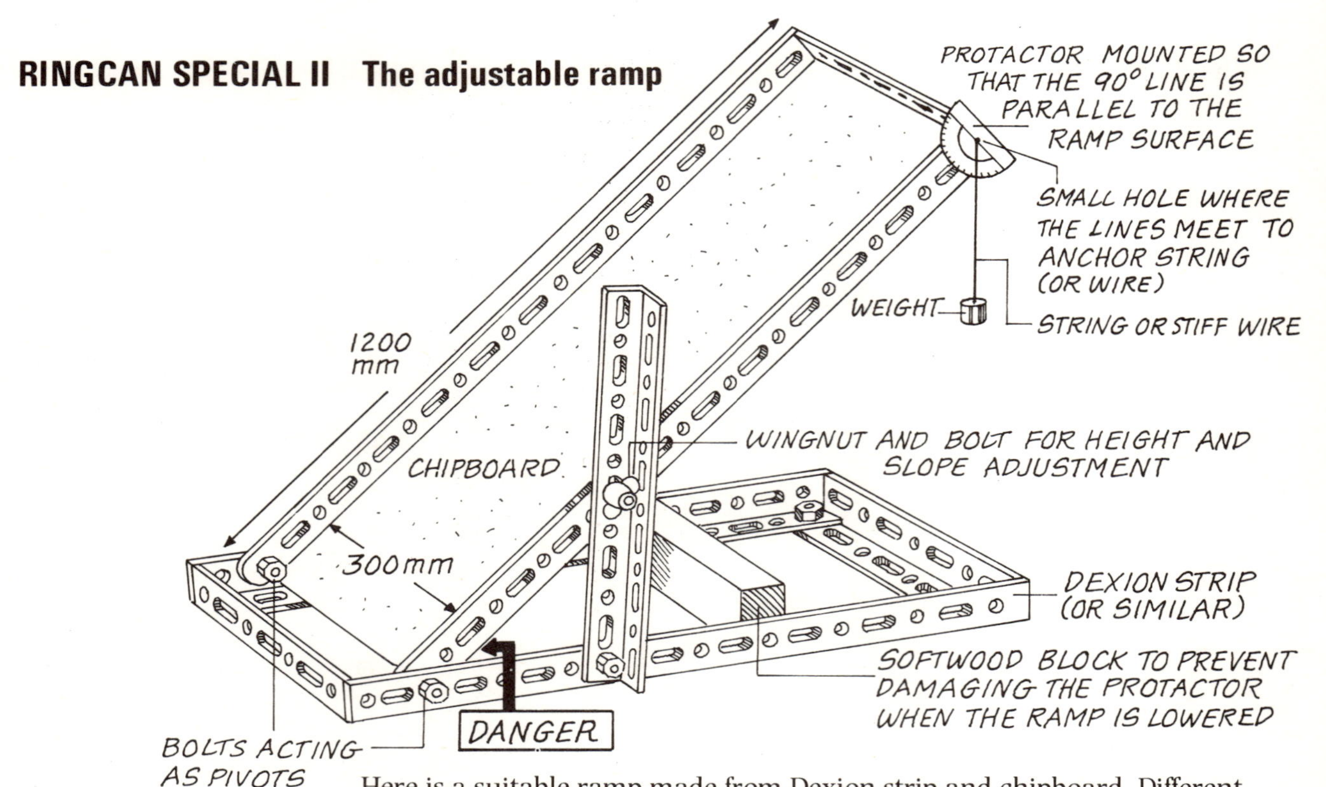

Here is a suitable ramp made from Dexion strip and chipboard. Different grades of abrasive paper can be glued at different positions along the ramp to study the effects of friction (see Using Science 4). The angle of the ramp can be measured directly by mounting the protractor and plumb line as shown. The angle of the slope is varied by using the strut and wingnut and bolt. Warn students of the danger to fingers in the gap between the sloping ramp and the base. The structure and materials are given only as a suggestion and may be modified to suit different ideas, needs or circumstances.

The obstacle course

The sketch is a possible layout for an obstacle course. Dimensions will depend on the size of the vehicles (length *and* width). The best solution is probably to present the design of the course as a mini Design Brief to the students concerned. They can jointly negotiate a design which will provide the fairest test for all of them. The course could be made from hardboard and softwood strip if a permanent version is required, or thick card and salvaged packaging materials if a cheap and temporary solution is more appropriate.

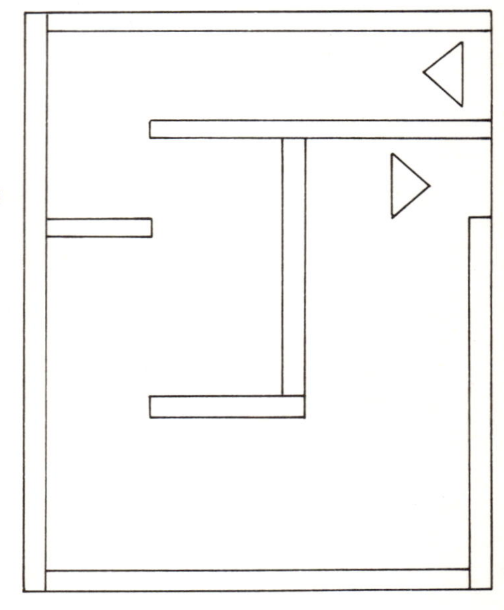

RESOURCES SUPPLEMENT 4

STEADY FREDDY

Do not bend the wire into its final shape until the whole of the steady hand game has been assembled and tested. The students need to drill all the holes in the acrylic before folding. Don't drill near the folds! The holes needed will usually be as follows:

a hole (about 2 mm) for each end of the welding rod (i.e. *two* holes);
a hole (3 mm) for the flexible wire to go through;
a hole (5 mm) for the LED (if required);
a hole for a fixing bolt to fasten the connector block directly underneath the acrylic top.

Don't forget to drill any holes needed for fixing the top to the bottom.
Fold the acrylic into shape and complete the case top. Put both pieces of sleeving onto the welding rod and thread it through the steady hand loop. The circuit, with the welding rod and flexible wire already fixed through the pre-drilled holes in the case, is now wired up. Test the circuit and, if all is well, bolt the connector block to the top through the hole provided. Test again and, if successful, bend the wire into shape.

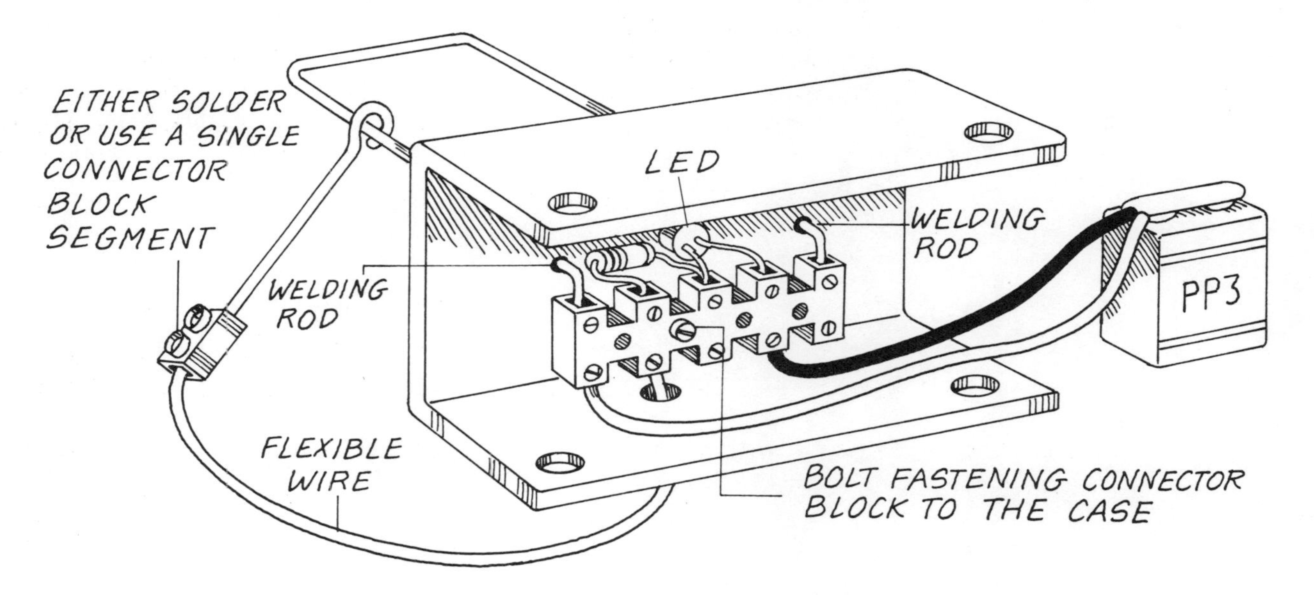

NOUGHTS AND CROSSES

Suggested basic materials for the box (per student)

Softwood, planed, 500 mm off 45 × 15 mm (finished size).
3-ply (4 mm) 2 off 125 × 125 mm.
The students can make grooves for sliding lids using a plough plane.

RESOURCES SUPPLEMENT 5

POCKET GAMES

You will need to make a Plug and Yoke mould as shown in Using Materials 6. The plug should be 90 mm diameter and the yoke hole 100 mm diameter. The acrylic should be no more than 3 mm thick and 150 × 150 mm square.

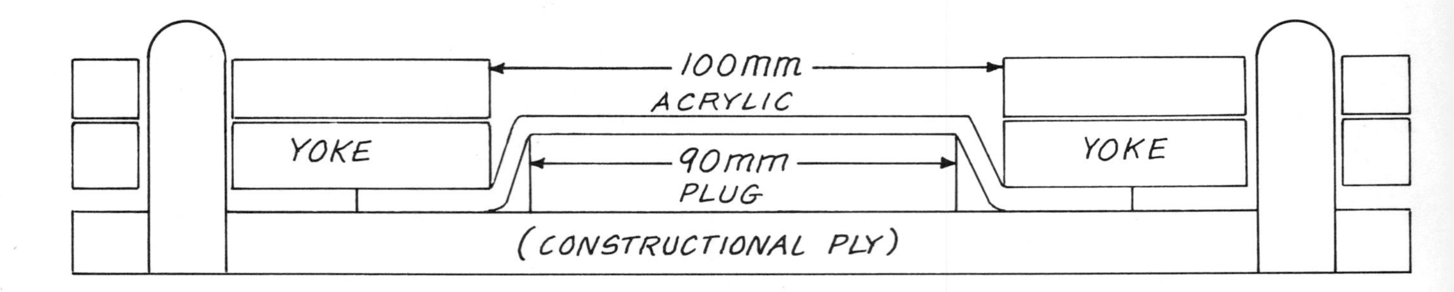

Once the dome and playing surfaces have been cemented together with Tensol, use a power jigsaw to cut the acrylic to no nearer than 6 mm from the edge of the dome (Using Materials 4). If this is not available, use a hacksaw to cut off the corners. Use a disc sander to tidy up the edge and remove saw marks. If no disc sander is available hold the game in soft jaws and use a file. When the edge is smooth, finish as described in Using Materials 6.

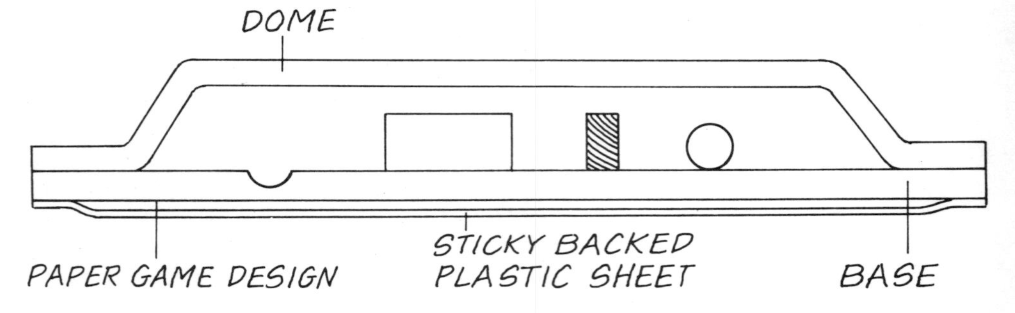

RESOURCES SUPPLEMENT 6

THE ELECTRONICS CIRCUIT BOARD

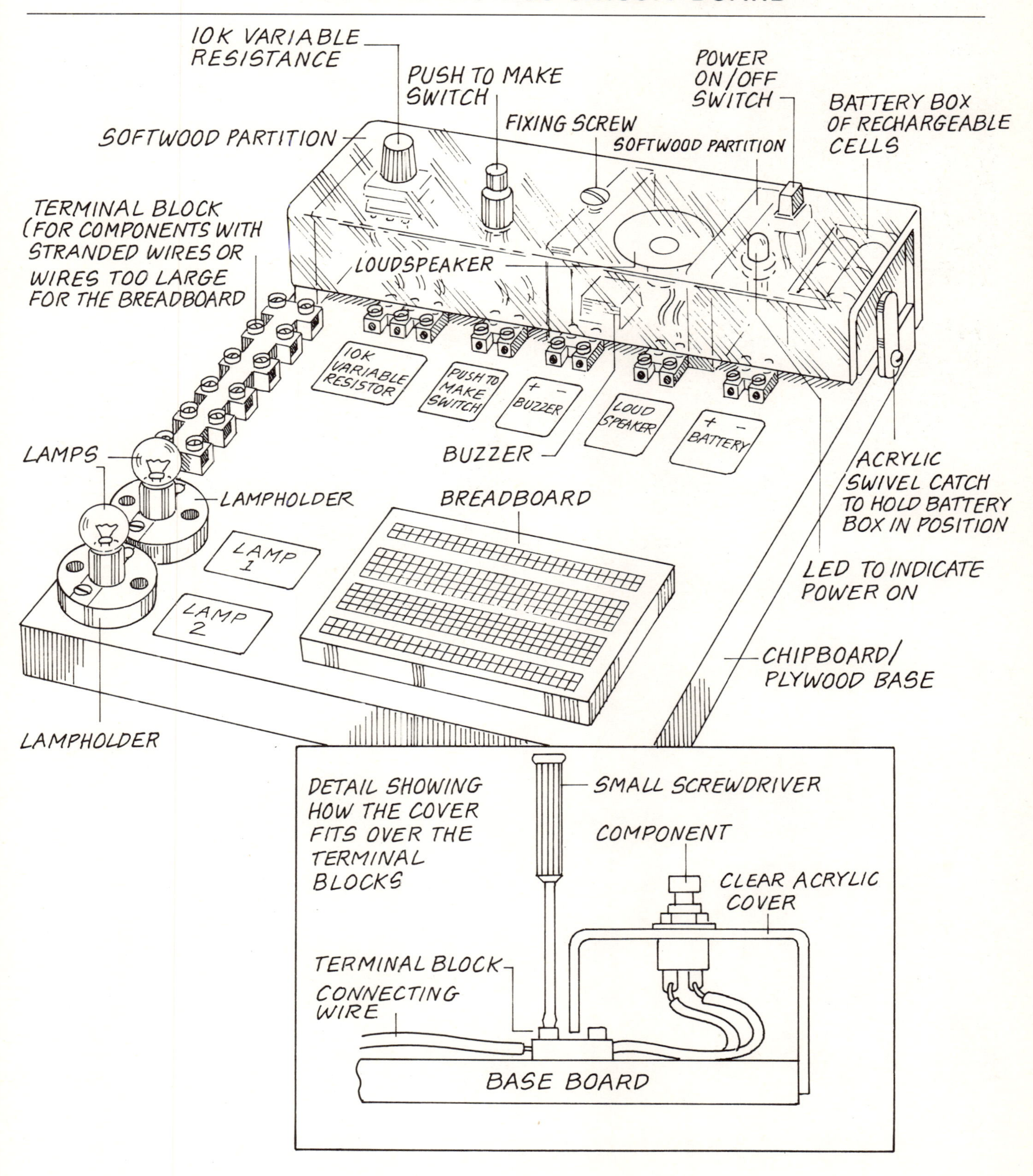

DESIGN BRIEFS IN USE

Investigation of structures – making a bridge

Investigation of mechanisms – levers and linkages

Investigation of electronics

Investigation of energy and control – transmission of power

Prototyping in waste materials – corrugated card

Graphics work

Making – using a power jig saw on hardboard

Making – using a disc sander on acrylic

Making – using a plug and yoke mould with acrylic

Finishing – smoothing and polishing acrylic

Testing a structure to destruction

Assessment/evaluation – presentation of the work and process

CDT FLOW DIAGRAM

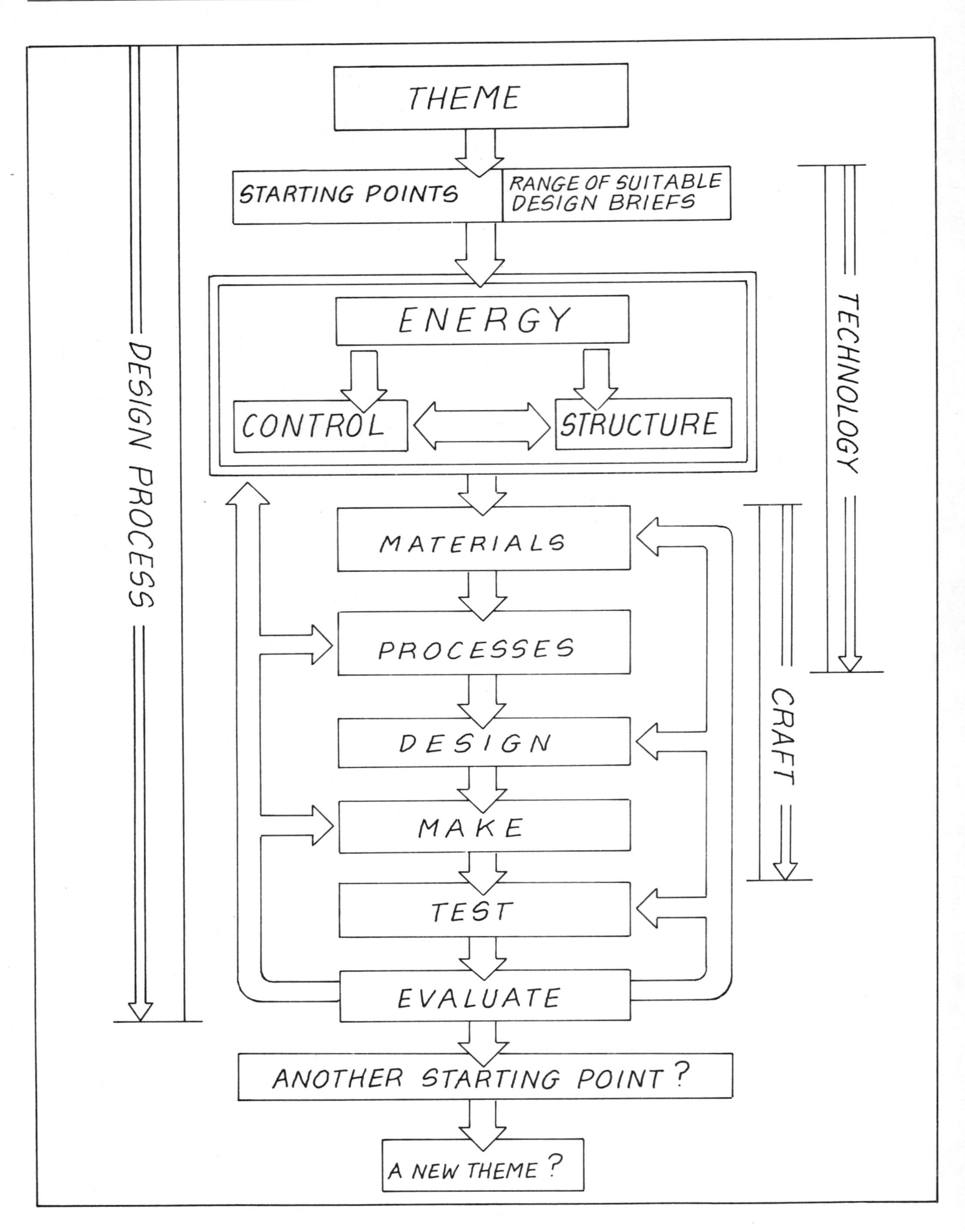

TEACHERS' PROJECT PLANNER

PROJECT

GROUP

STARTED

CHECKLIST

1. Have I organised all the visual aids, examples, samples of the materials, I need to introduce this project?
2. Are all the materials and equipment required for planning and prototyping ready? Where necessary, have they been cut to size?
3. Have I ordered any special items not in stock (e.g. electronics components)?
4. Have any jigs, moulds or templates to be made?
5. Are all the materials needed for making the project ready? Are they cut to size, where appropriate?
6. Are all the necessary tools, equipment or machines ready for use (e.g. drill sizes, correct blade in jigsaw)?
7. Have I remembered all aspects of safety in the workshop relevant to this project?
8. Are there any pieces of test equipment to be made or supplied?
9. Do I know where the incomplete work will be stored?
10. Has all the work been clearly labelled?
11. Has any child been absent and missed an important part of the project?
12. Have all the children completed an evaluation of their work?
13. Have I completed an evaluation of their work?
14. Have I evaluated the way this project has developed?

COMPLETED

POINTS FOR NEXT TIME

BST 1 PAPER STRUCTURES

DESIGN BRIEF

1 Using the strip of paper provided (75 × 510 mm), design and make a bridge which will support a load greater than 100 g while spanning a gap 300 mm wide.

2 Using three strips of paper (75 × 510 mm), design and make a bridge spanning a 300 mm gap and capable of supporting more than 500 g.

3 Using as small a number of strips as possible, design and make a bridge capable of spanning a 300 mm gap which will support a weight of more than 1.5 kg.

POINTS TO CONSIDER

If you use glue, remember to let your bridge dry before testing.

Your bridge will be stronger if it is well made – tears in your paper will make it weaker!

Put your results down in table form so you can compare one design with another.

THE TEST

To test your bridges a test station will be set up as shown on the right. You are not allowed to fasten your bridge onto the blocks in any way. Your bridge must rest on the top. The weights must be in the middle as shown in the diagram.

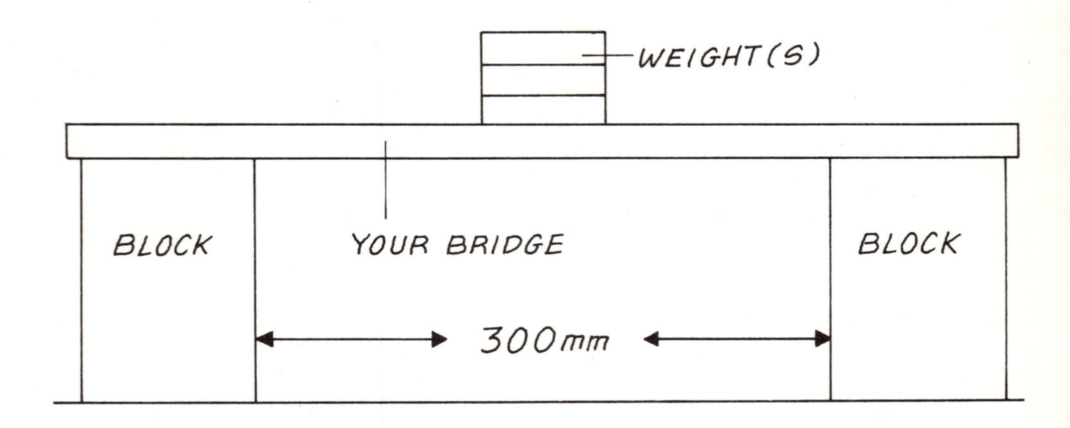

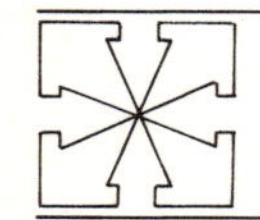

Make a list of all the information sheets you have used.
List any other sources used.

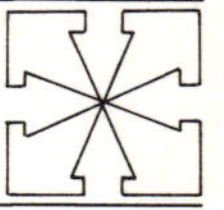

BST 2 PAPER STRUCTURES

DESIGN BRIEF Using an A4 sheet of cartridge paper and PVA adhesive, design and make a pillar which will support a weight of at least 5 kg at a height of 100 mm above the table top.

POINTS TO CONSIDER How can you make your paper more rigid?
What shapes do you know which show great rigidity?
Always put the weights on carefully – do not drop them.
Make a table so you can compare your designs.

THE TEST

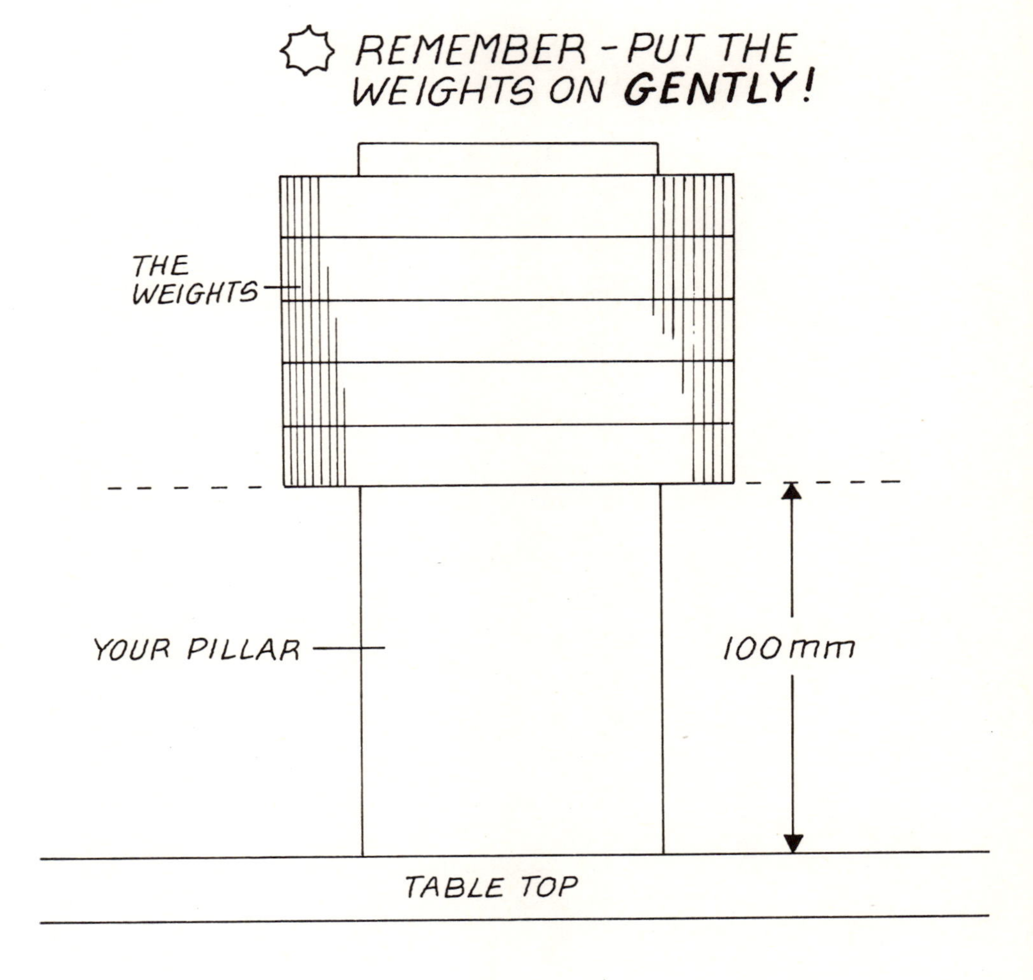

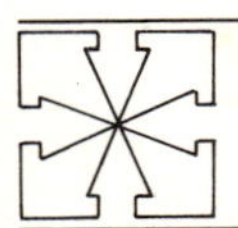

**Make a list of all the information sheets you have used.
List any other sources used.**

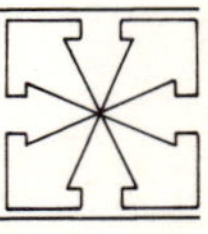

BST 3 PAPER STRUCTURES

DESIGN BRIEF

1 From the two paper art straws given, use one of the straws to make a base which will make the other straw able to stand upright.

2 Using an A3 sheet of paper and 1 m of gummed paper tape (25 mm wide), design and make a free-standing tower. It must be able to support a 20 g weight at the greatest height above its base.

POINTS TO CONSIDER

To be free standing and to remain upright, your tower needs a good base.
In 2, the more paper you use for the base the less there is for the tower.

THE TEST

Free standing means your tower must stand up on its own without any help from you or anything else. It must support the weight for at least 30 seconds.

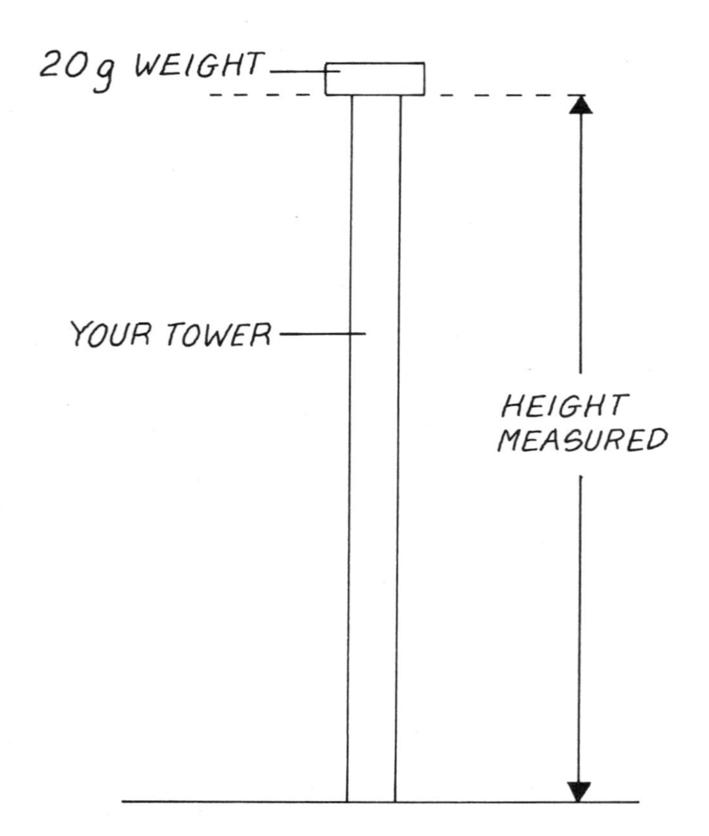

Make a list of all the information sheets you have used.
List any other sources used.

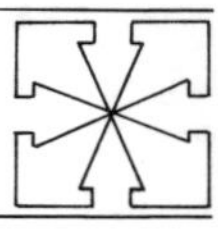

BST 4 STRAWMOBILE

DESIGN BRIEF Using only the materials provided (which are 4 art straws, 2 cocktail sticks, stiff card for 4 wheels (or ready made wheels), 1 small piece of card and any necessary glue(s)) make a vehicle which will carry up to 300 g as far as possible in a straight line across the workshop floor using the ramp supplied.

POINTS TO CONSIDER The axles must be exactly in line (parallel) if the vehicle is to run straight.
The vehicle must run smoothly and not bob up and down or wobble as this wastes energy which could be making it go further.
Where does the energy come from?

BASIC CHASSIS

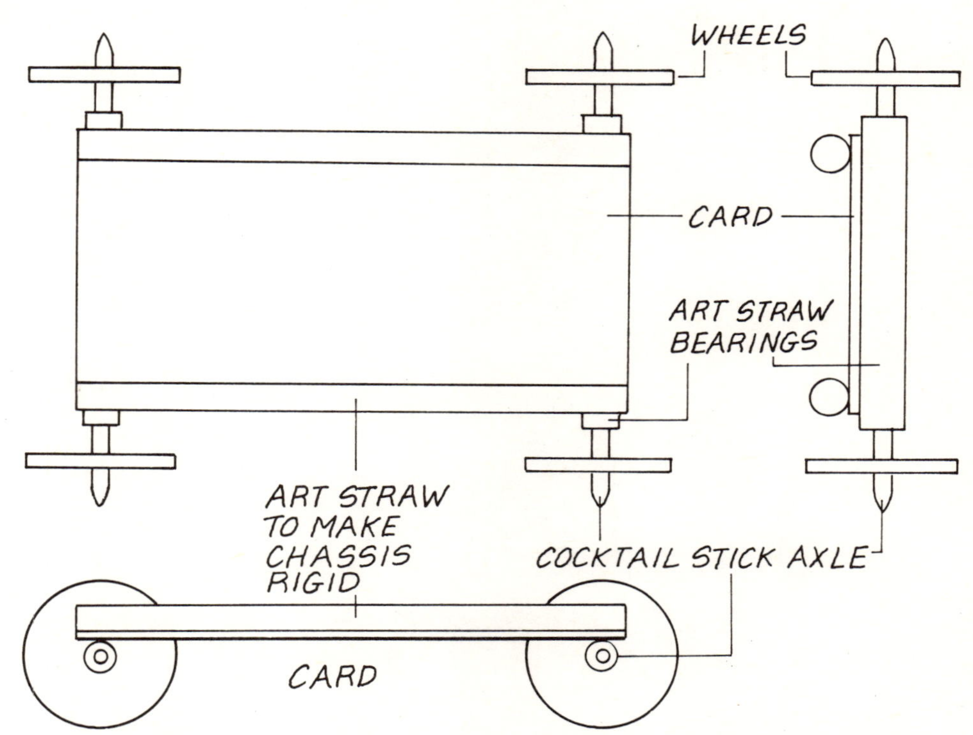

In this view the two near wheels have been removed to show the chassis more clearly.

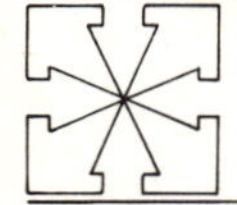

Make a list of all the information sheets you have used.
List any other sources used.

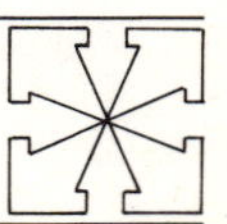

BST 5 EGGTOR PROTECTOR

DESIGN BRIEF Design and make a box which will carry an egg safely through the post. The box must be decorated to make it attractive and have an address label as part of the design.

The box itself and all internal parts must be made from a piece of thin card no larger than 315 × 520 mm.

POINTS TO CONSIDER Is an egg equally strong in all directions?

You should make your box so the egg can easily be put in and taken out.

Your egg box, if successful, will be made in large numbers. Is it easy to make?

How would you advertise your egg box?

THE TEST Your box should protect an egg dropped from a height of 1 m onto a hard surface. To save on eggs start at a much smaller height first and work up to 1 m!

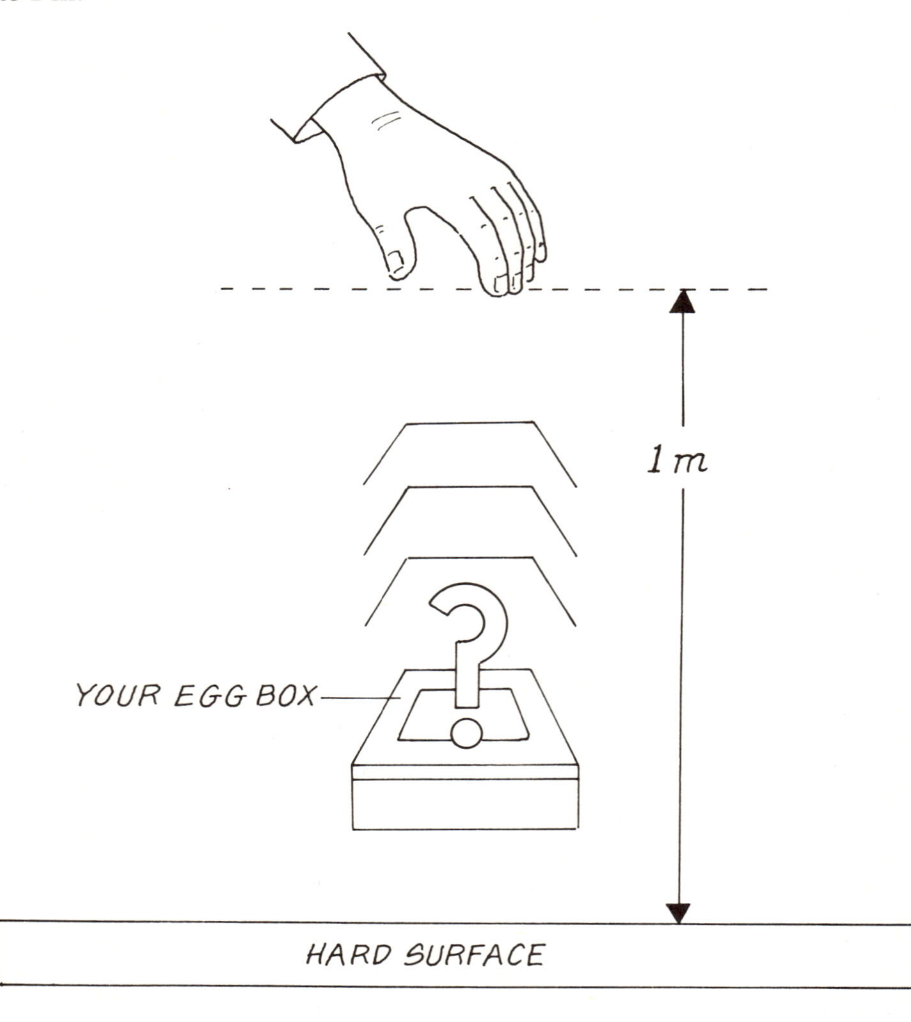

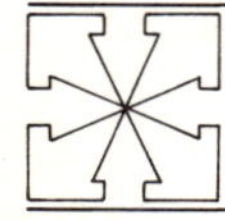

Make a list of all the information sheets you have used.
List any other sources used.

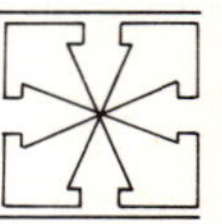

BME 1 LOOPY LINKS

DESIGN BRIEF Using only the materials provided (i.e. hardboard 300 × 150 × 4 mm, any 6 mm dowel needed, an elastic band and plastic tubing to fit the dowel) design and make a funny face, with or without body, which has parts that move when a lever is pulled.

POINTS TO CONSIDER Try out some simple mechanisms (e.g. levers and linkages) in stiff cardboard. Using a piece of squared paper which is the size of the hardboard, plan how you will use the board to make the face and *all* other parts. Note that parts with pivots must be 20 mm wide for drilling. Now make a working model (prototype) in stiff card of your idea using paper rivets as pivots. When your idea is working, take your model apart and draw round the pieces on some white paper the same size as the hardboard. Cover the hardboard thinly with PVA – use a scrap of card as a spreader – and stick the paper down flat, i.e. without air bubbles. Leave it to dry, ideally overnight. Cut out the shapes; drill and put them together. Colour the paper and spray with clear lacquer.

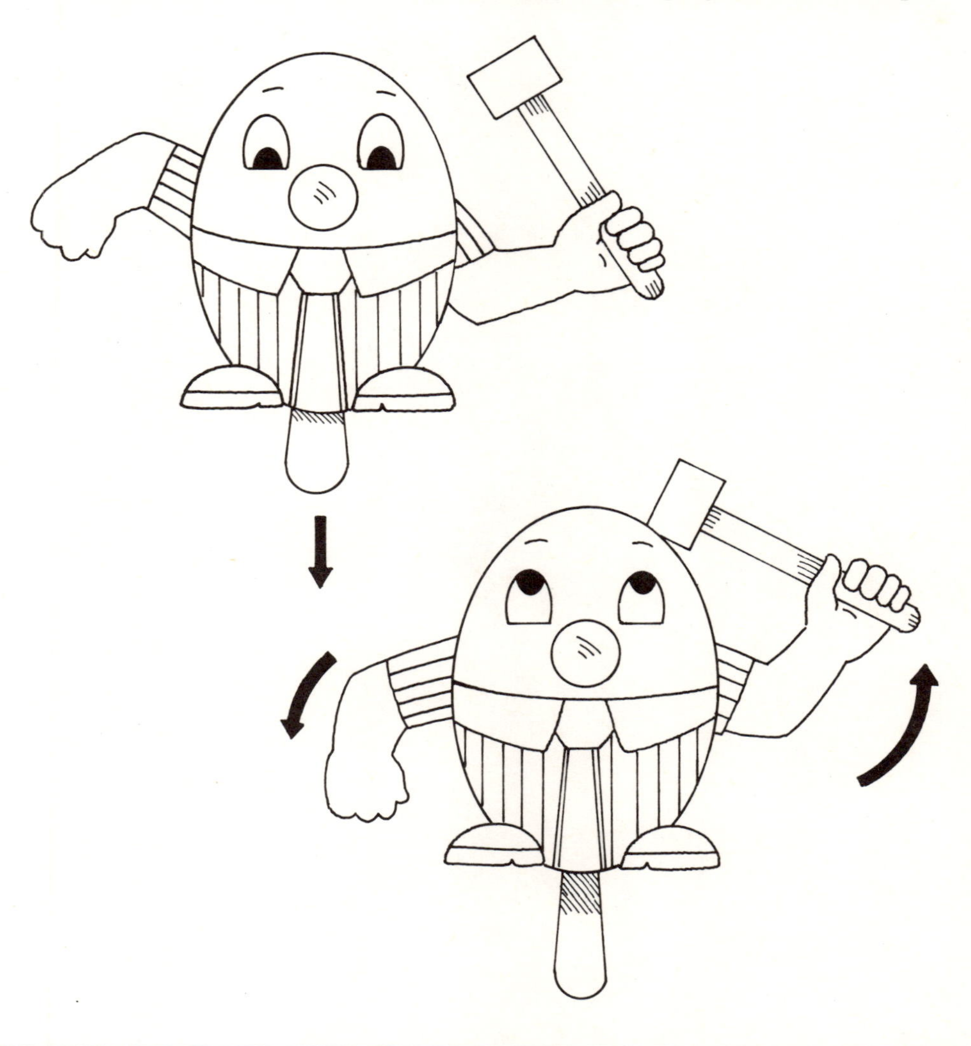

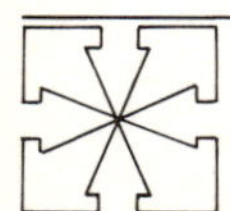

Make a list of all the information sheets you have used.
List any other sources used.

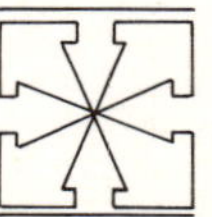

BME 2 MAKE A SCENE

DESIGN BRIEF

Many museums and similar places have some exhibits where you press a button and either something lights up or moves. With some displays, several things happen one after the other; this is called a sequence.

Design and make a scene in which something very amusing happens when a button or lever is pressed. Your scene can be mechanical or use electronics or both. It could make strange noises. Your device should be able to fit inside a box which is 100 × 150 × 200 mm.

POINTS TO CONSIDER

How can you make several things happen one after the other (i.e. create a sequence)?

Could you make lights flash?

How might your device produce sounds?

Your scene should be attractive, well-made and finished.

Design a thin card display box suitable for showing your device in a shop.

Create an advertising jingle for your device.

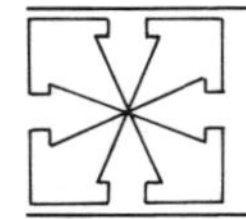

Make a list of all the information sheets you have used. List any other sources used.

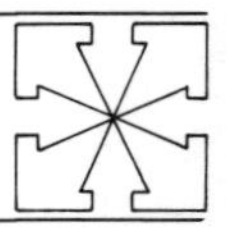

BME 3 PUSH–PULL TOY

DESIGN BRIEF You may have seen pull- or push-along toys which have parts that move as the toy is pulled or pushed along. Design and make such a toy suitable for a young child. Choose carefully from the materials you have available.

POINTS TO CONSIDER The toy must be safe to use. This means there should be no sharp points or edges even if the toy is broken. Why must you avoid using nails?

If you use any paint or varnish, it must not be poisonous. Can you suggest why?

The toy should be amusing and attractively finished so that it catches the eye.

You may find it useful to find out about levers, cranks and cams. What mechanism could you use in this case?

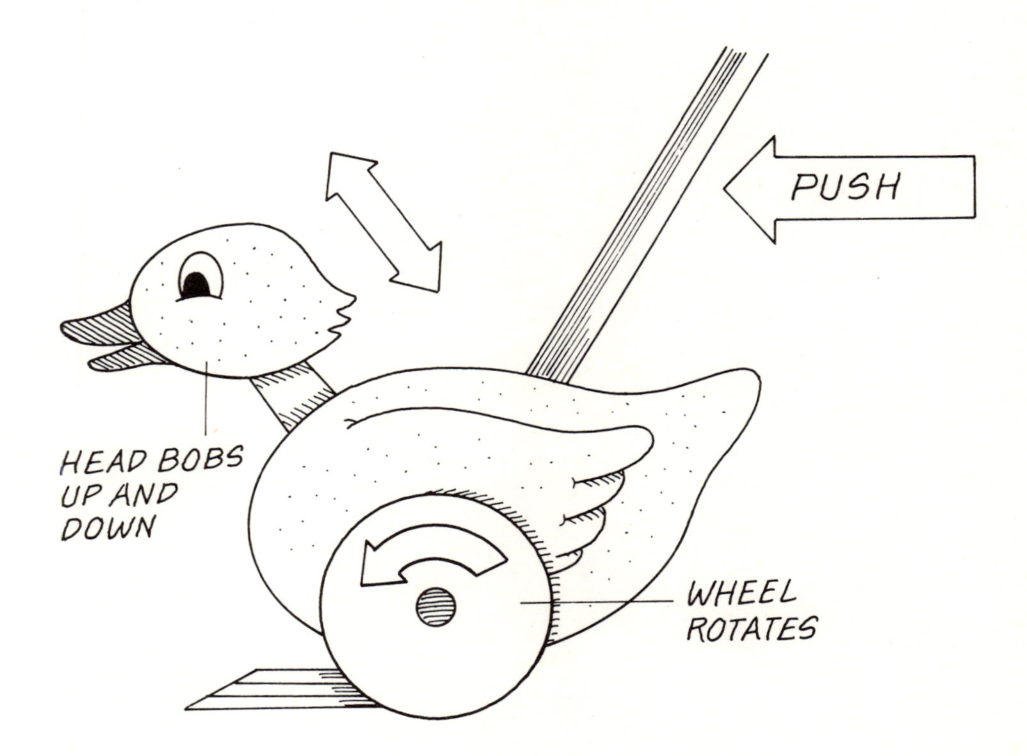

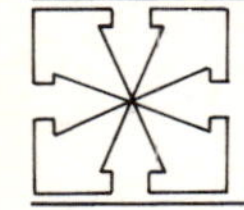

Make a list of all the information sheets you have used. List any other sources used.

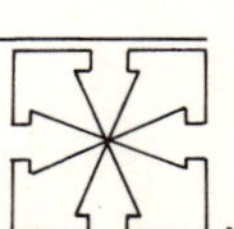

BME 4 PINBALL CRAZY

DESIGN BRIEF You have probably seen pinball machines in arcades. You may also have heard of a much older game called bagatelle. Design and make a pinball game using a marble or ball bearing. The base should be made from a piece of hardboard 175 × 250 mm. You can make the pockets, obstacles, deflectors, etc., from softwood, dowel, panel pins, elastic bands, or plywood. The top can be made from clear 3 mm acrylic sheet and held by five small woodscrews.

POINTS TO CONSIDER Your board must be sloped so it can work properly by using microswitches or electrical contacts. (Remember: a ball bearing conducts electricity, a marble does not!) You could make lights (i.e. LEDs or lamps) flash or a buzzer sound as the ball strikes. You could use levers and linkages to move 'flippers' to nudge the ball in the right direction as it rolls by. Below are suggestions for propelling the ball.

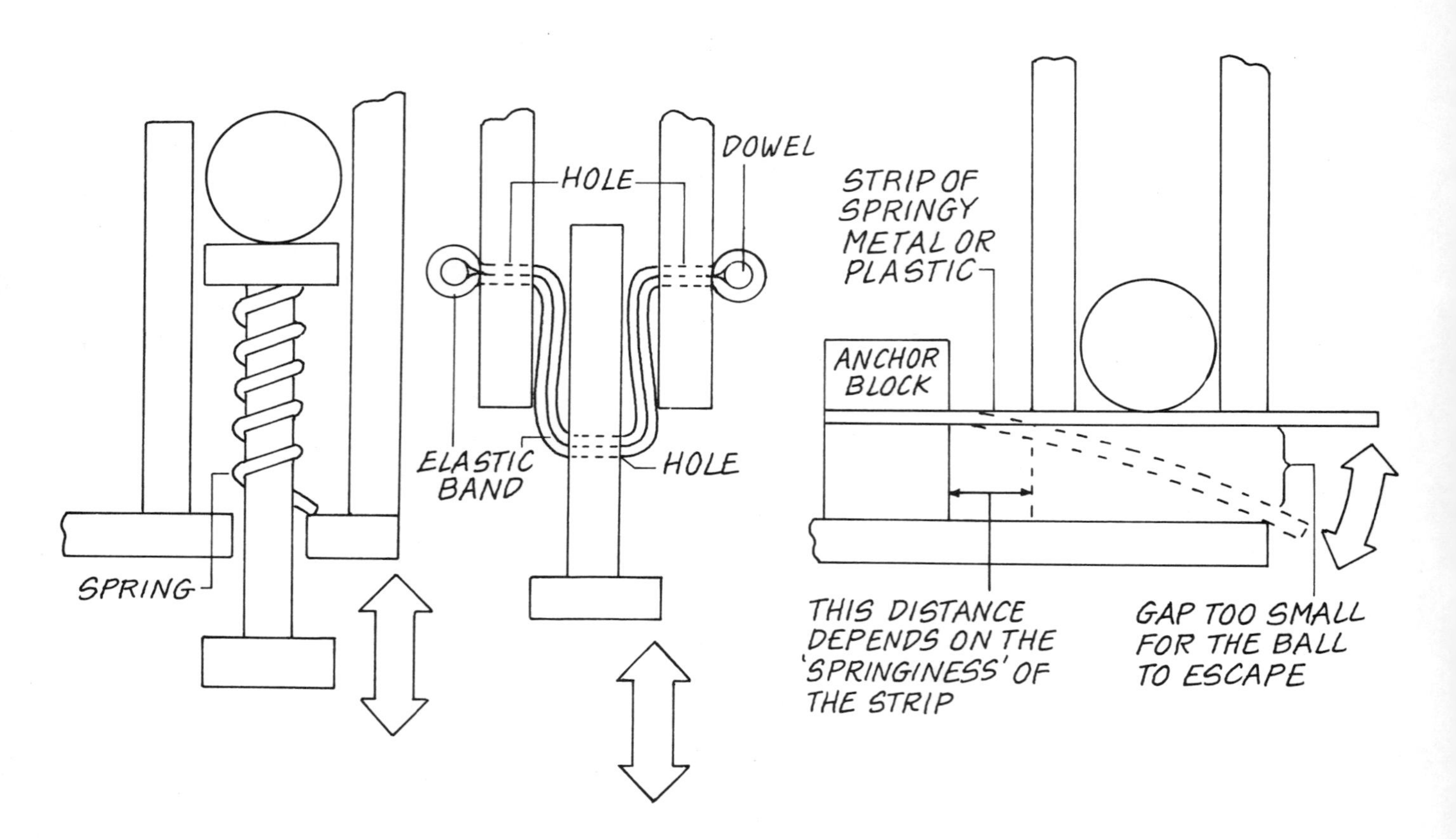

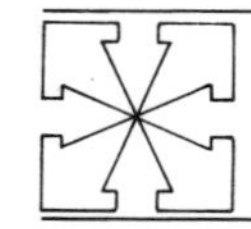

Make a list of all the information sheets you have used.
List any other sources used.

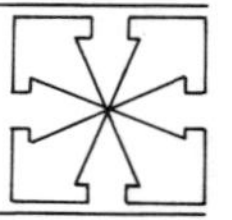

BME 5 CATCH IN THE BOX

DESIGN BRIEF Design and make a small decorative lidded box in wood or acrylic with a secret catch. It must be no larger than 45 × 45 × 100 mm. The wooden box may be made by drilling a 30 mm diameter hole in 45 mm square-section wood.

POINTS TO CONSIDER How will you disguise your catch to make it secret:

1 by making it appear to be part of the decoration?
2 by hiding it in an unusual position?
3 by requiring some sort of key?
4 some other way?

You will need to design and make experimental catches before you design your box!
Here are some not very secret ideas!

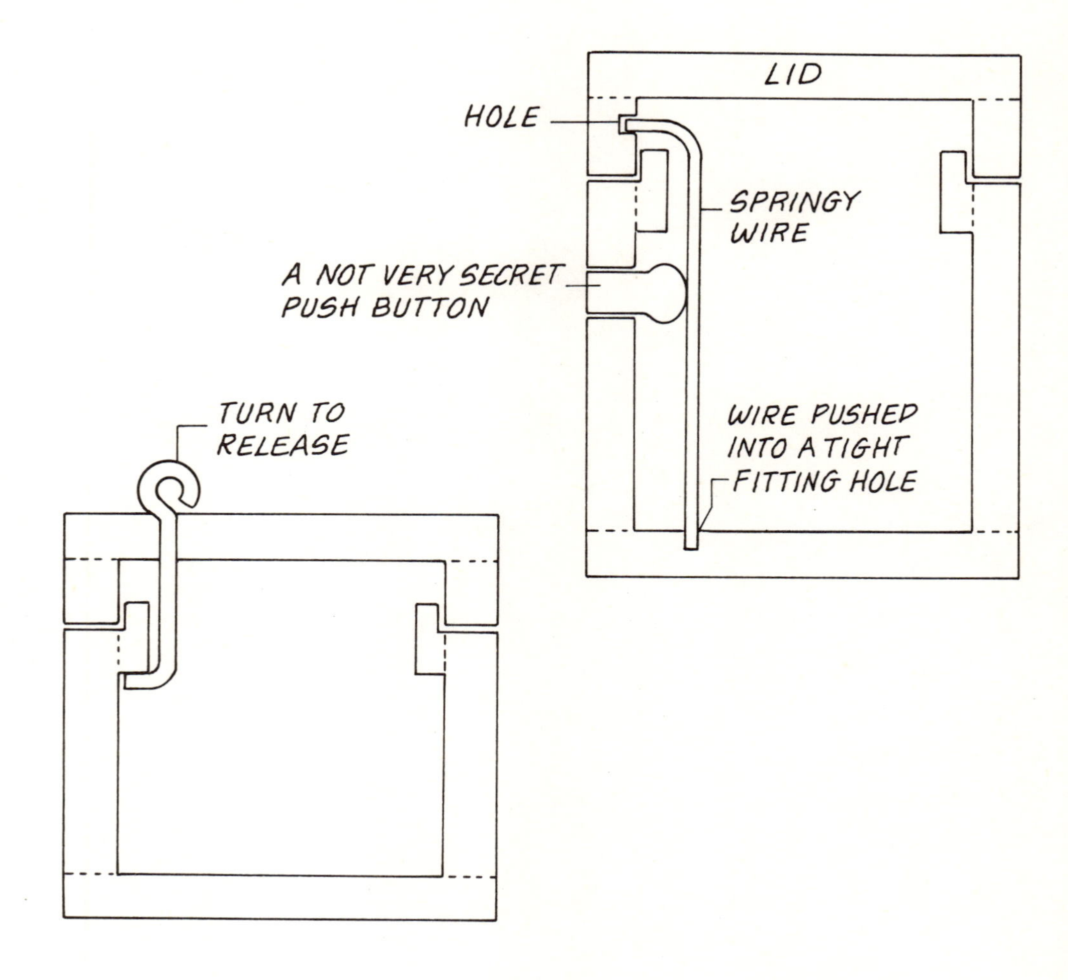

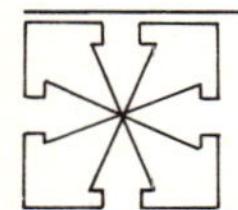

Make a list of all the information sheets you have used.
List any other sources used.

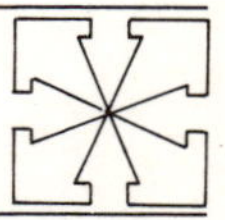

BSC 1 MARBLOUS BOATS

DESIGN BRIEF

From a piece of aluminium kitchen foil which is 150 × 150 mm square, design and make a boat which will carry the largest cargo of marbles for 20 seconds without sinking.

POINTS TO CONSIDER

The flat sheet of foil will float on the surface of the water. If you gently push the foil under the water, taking care not to trap air underneath, it will slowly sink to the bottom. Can you explain why this should happen?

If you crumple the foil up in air as tightly as you can and drop it in the water, it still floats. If you crumple the foil under water it will sink. Why?

If you use a 30 g lump of plasticine for your boat, how big a cargo of marbles will it carry? Try using paper instead of foil. What extra problems does this cause?

THE TEST

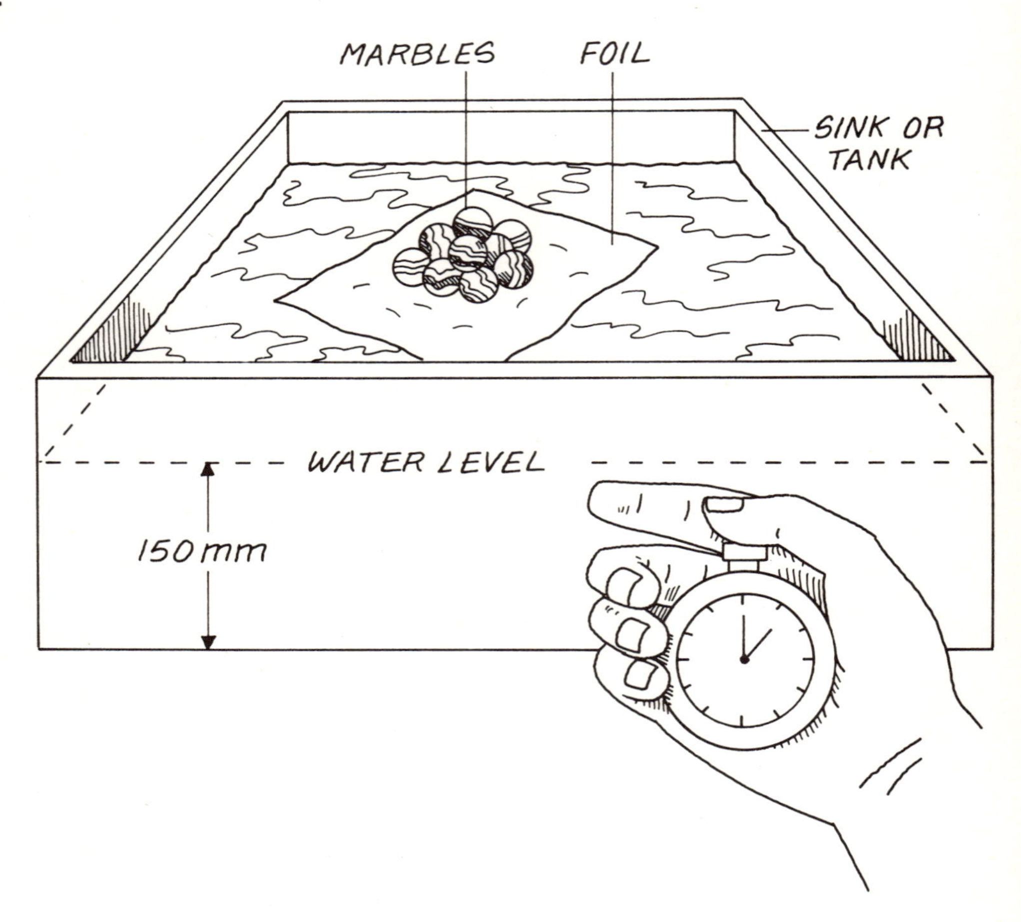

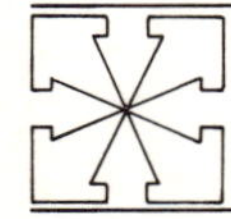

Make a list of all the information sheets you have used. List any other sources used.

BSC 2 TOPPING IDEAS

DESIGN BRIEF Using only the card and the cocktail stick or thin dowel provided, design and make 'the best top in the Universe'.

POINTS TO CONSIDER What do you think is likely to be the best shape?

Does the size of the top make a difference?

Does the position of the cardboard on the stick make a difference?

How can you make your top exciting to look at:

1 when it is standing still?
2 when it is moving?

THE TEST Is the best top the one that spins the fastest? If so how can you count the number of turns per second?

Is the best top the one that spins the longest? You can test this with a friend using a stop watch.

Is one 'go' a fair test or is it better to take the average of several 'goes'?

Is the best top the most attractive when it is:

1 standing still?
2 moving?
3 both?

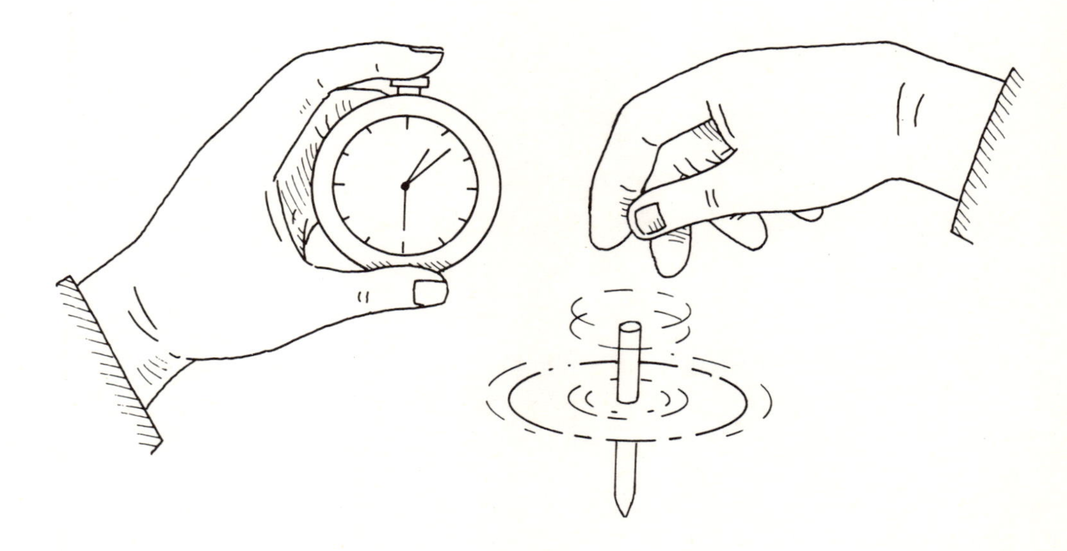

Are there any other ways in which tops can be best?
How would you test them fairly?

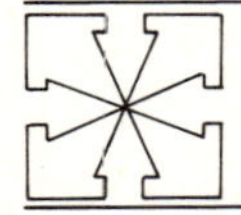

**Make a list of all the information sheets you have used.
List any other sources used.**

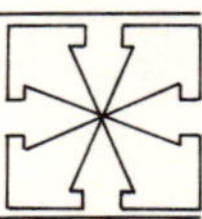

BSC 3 A4 FLYING WING

DESIGN BRIEF From an A4 sheet of paper make a flying wing as shown below. First make it fly straight, then perform loops, turns and rolls. To do this you will need to experiment with wing flaps, stabiliser fins and rudder. Now design and make your own plane using up to two A4 sheets of paper so that it will fly straight or do 'stunts' whenever you choose.

POINTS TO CONSIDER How will you make your plane look attractive? Use scissors to cut your 'wing' to look like a bird or an aeroplane.

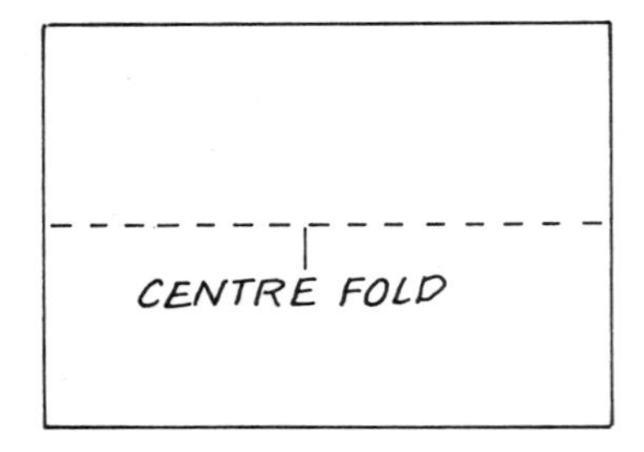

1 Fold in half and then open out again.

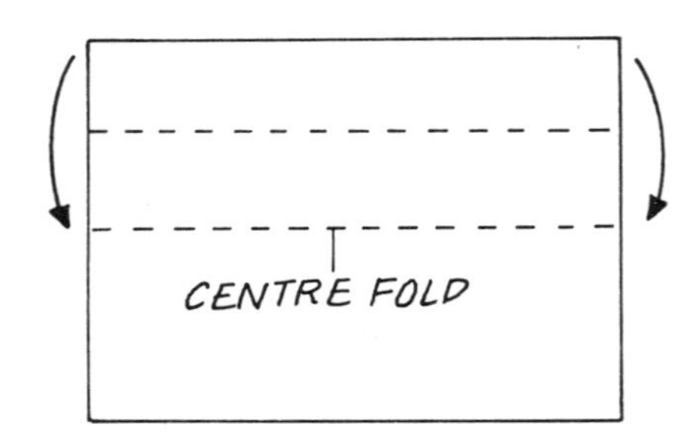

2 Fold this half in half (long edge to middle).

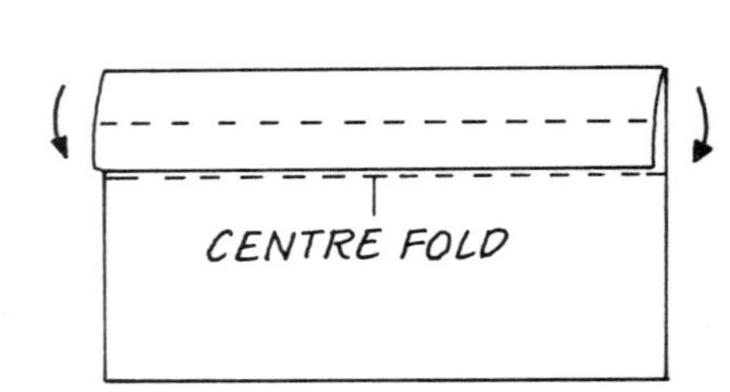

3 Fold in half again (folded edge to the centre fold).

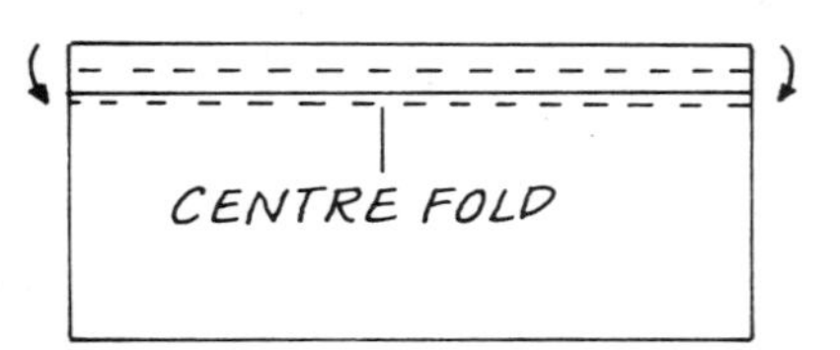

4 And fold again (folded edge to the centre fold again).

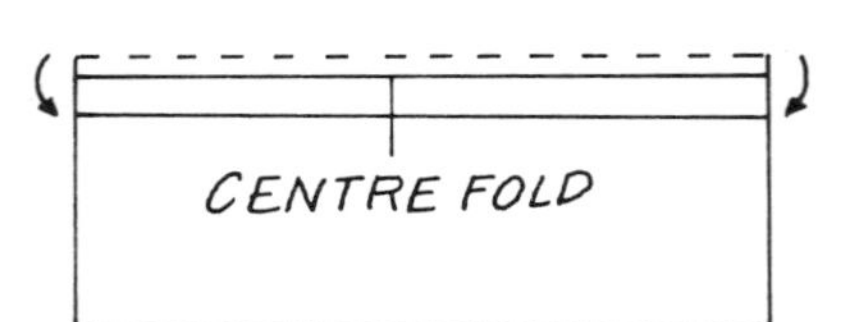

5 Fold the whole of the folded half on top of the unfolded half.

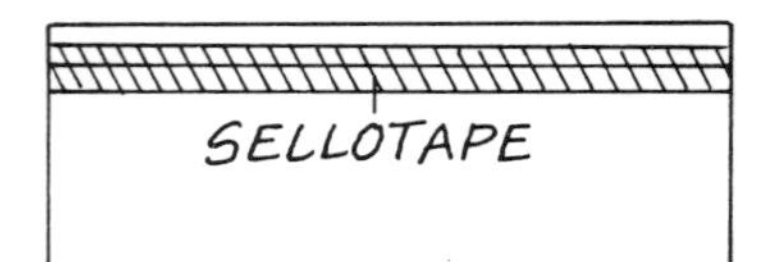

6 Sellotape the folded half into position.

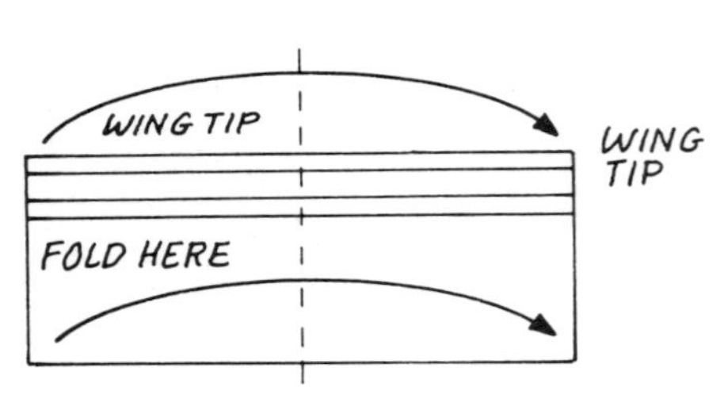

7 Fold in half, wingtip to wingtip, and open out again.

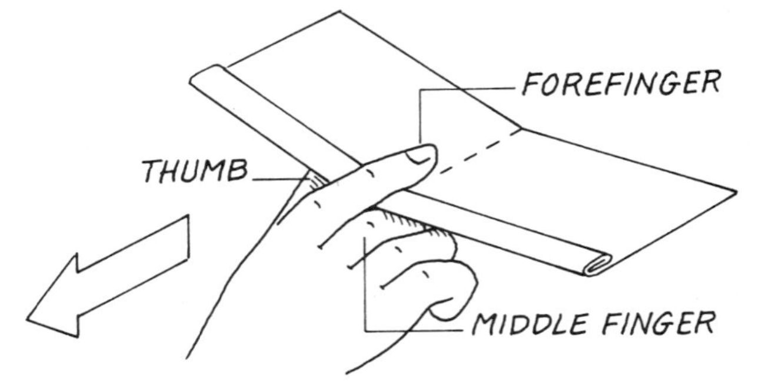

8 Launch 'backhanded' away from you like this.

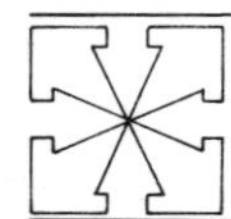

Make a list of all the information sheets you have used.
List any other sources used.

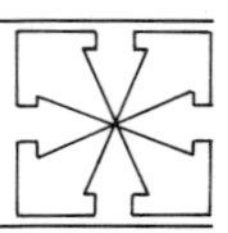

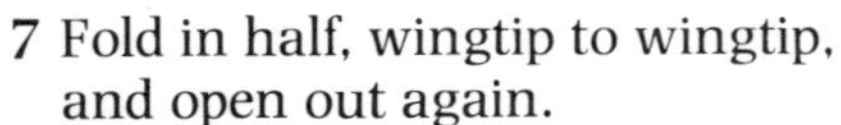

BSC 4 THE SIMPLI-KITE

1 Cut two lengths of 6 mm dowel, one 700 mm long and the other 560 mm long.

2 Cut a notch 5 mm deep at each end of the two dowels with a saw.

3 Using a small round file, make a shallow groove across the middle of the shorter dowel. This must be as accurate as possible.

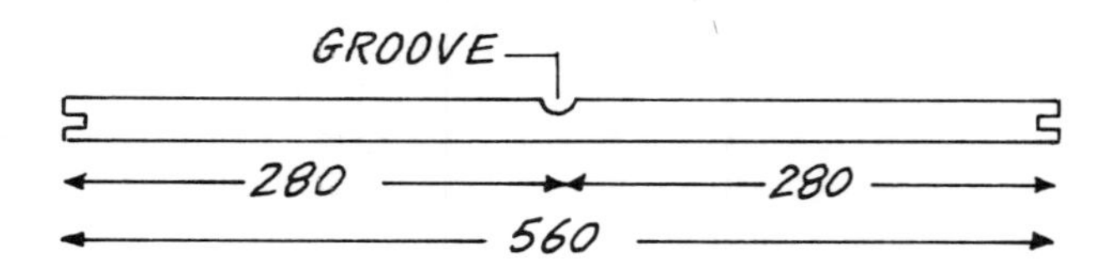

4 Make a shallow groove on the longer piece 150 mm in from one end.

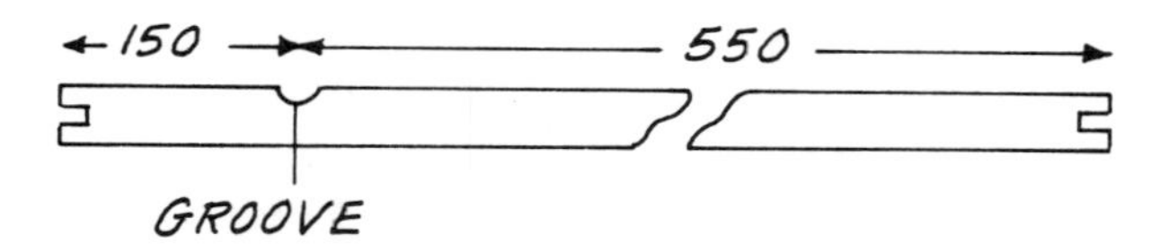

5 Glue the grooves together and lash tightly with string round the joint. Tie and anchor the string with a thin coat of glue.

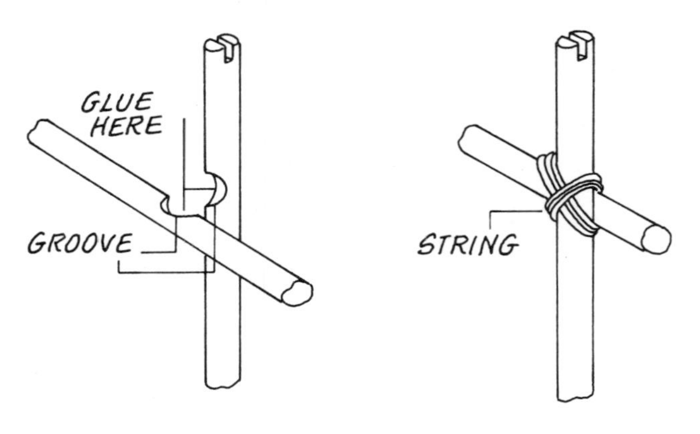

6 Take a length of string and slip it into each notch in turn at the ends of the dowels. Pull it tight, knot and glue.

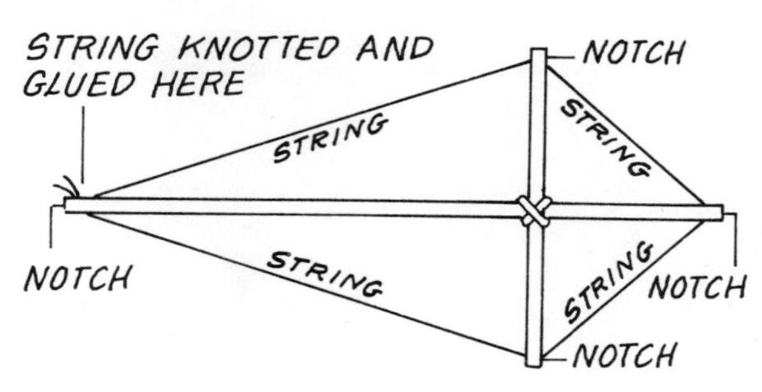

7 Cut a piece of polythene bin liner to allow a 20 mm hem all round. Fold the hem over the string and stick in position with sellotape. (A partner is useful here.)

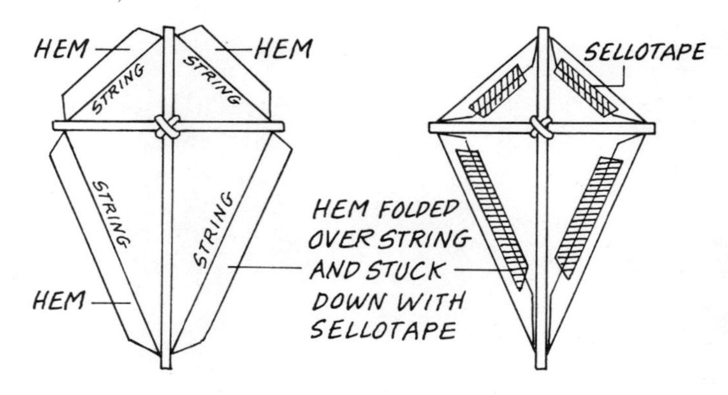

8 Make two holes in the polythene just above where the dowels cross and tie the top bridle string through these holes to the upright.

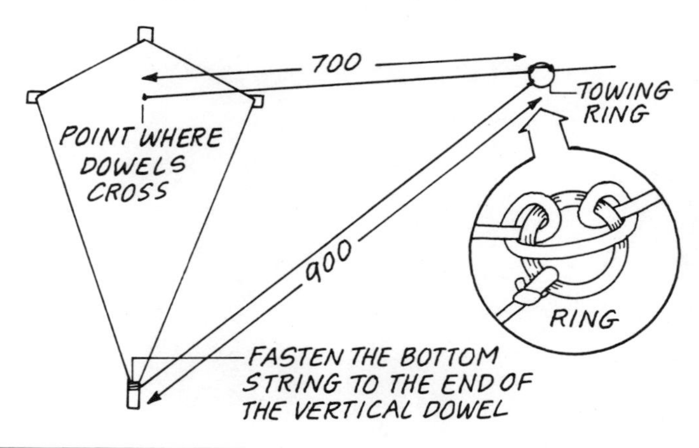

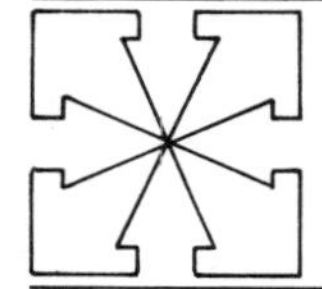

Materials 6 mm dowel 1260 mm long
Polythene sheet from bin liner 700×560 mm
Small split ring Thin string Glue and sellotape

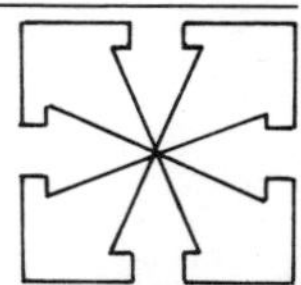

BSC 5 GO FLY A KITE

DESIGN BRIEF

Make the Simpli-kite as shown on the instruction sheet. Try to make your kite easier to control by:

1 adding a tail – try different lengths until you find the best;
2 altering or adjusting the bridle to change the angle at which the wind strikes the kite.

Successful kites can be made in a whole variety of shapes.

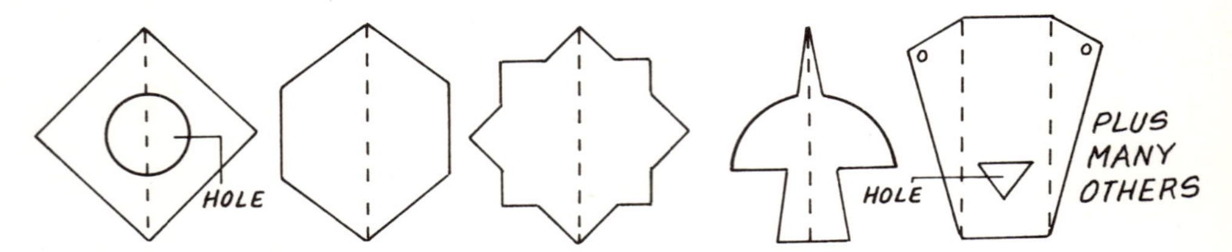

POINTS TO CONSIDER

They are all symmetrical about the vertical.
Design and make your own kite which should be: **1** no larger than 700 × 700 mm, **2** fly well, **3** be easy to control, **4** look interesting in flight. You may use any of the following materials: dowel or split bamboo, polythene sheet, lightweight cloth or strong paper, string and adhesives.

If you use cloth you could decorate your kite using tye-and-dye or batik. Polythene sheet is quite difficult to decorate. You could use coloured sticky tape. Many kites from China are shaped and decorated to look like birds or butterflies. How will you decorate yours?

To launch your kite do not gallop across the field dragging it behind you. You will only succeed in damaging your kite. Get a friend to help you by holding the kite above their head several metres downwind from you. As your friend releases the kite, move backwards until the kite rises, then slowly let out more line. **Danger:** avoid overhead power cables!

To make a tail, tie bunches of folded paper at intervals along a string fixed to the bottom of your kite. Tails are usually five or six times the length of the kite.

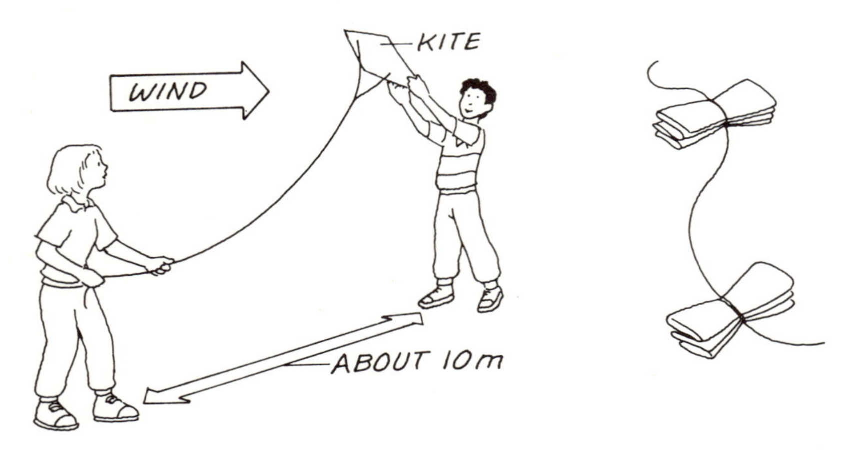

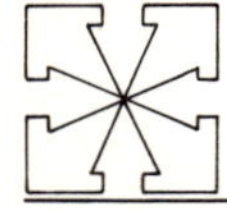

**Make a list of all the information sheets you have used.
List any other sources used.**

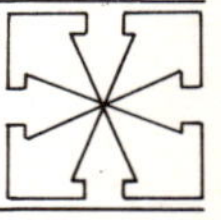

BSC 6 WEIGHING IT UP

DESIGN BRIEF

Find as many different ways of weighing as you can and make notes and sketches to show how they work.

Using as little material as possible, design, make and test a simple device for weighing accurately in the range of 0–100 g or 0–500 g. You will need to experiment before designing your device.

Here are some suitable materials: wood (softwood, plywood, dowel), acrylic sheet (or other plastics), packaging materials (e.g. margarine tubs), elastic band or small spring, wire, stiff card, gear or pulley wheels.

POINTS TO CONSIDER

How are you going to make an accurate scale for your machine (i.e. calibration)?

Find out how you could use gears or pulleys to make your balance more sensitive.

If you are weighing foodstuffs, then easy-to-clean hygienic surfaces are very important. For this reason there should be no rough edges or surfaces where germs could breed.

Your work should be finished to a high standard and should be attractively styled.

THE TEST

You will be asked to weigh a mystery object which has been checked against a very accurate chemical balance. The device which gives the weight nearest to this will be judged the most accurate.

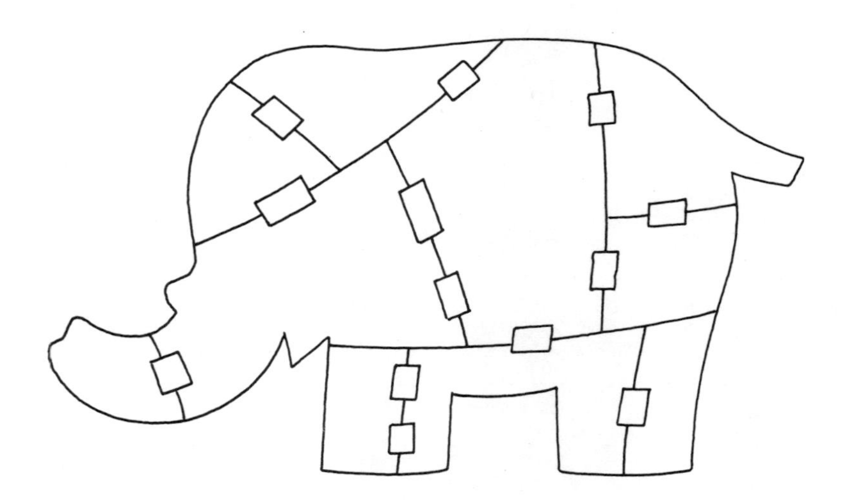

THE MYSTERY OBJECT?

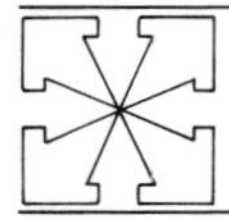

**Make a list of all the information sheets you have used.
List any other sources used.**

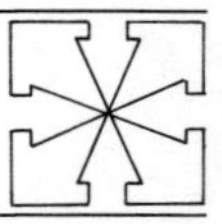

BEC 1 HOT RUBBER

DESIGN BRIEF Design, make and test a hot rod powered by a motor made as shown below. You will need the following materials:

A4 sheet of thin card
2 axles
materials for motor
4 wheels
propeller

POINTS TO CONSIDER Your hot rod should run straight.
The weight of your vehicle is very important.
Try the body shape in paper first (prototype).
Your vehicle may not be the fastest, but how can you make it look fast?

THE TEST This is the fastest 'rod' over 1.5 m.
It is found by measuring the time taken to cover that distance.
Take the average of five runs.

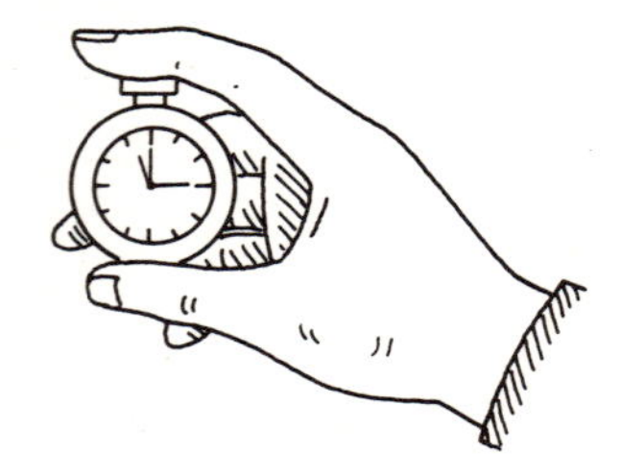

HOW TO MAKE THE MOTOR First open out a paper clip as shown. Thread a washer and bead onto the clip. (You could use a washing-up liquid spout instead.) Now add the propeller. Hook on the elastic band. Pull the other end of the elastic through the tube with a hooked piece of wire. Anchor the elastic with a 2.4 mm thick metal rod fitted into a notch in the tube as shown.

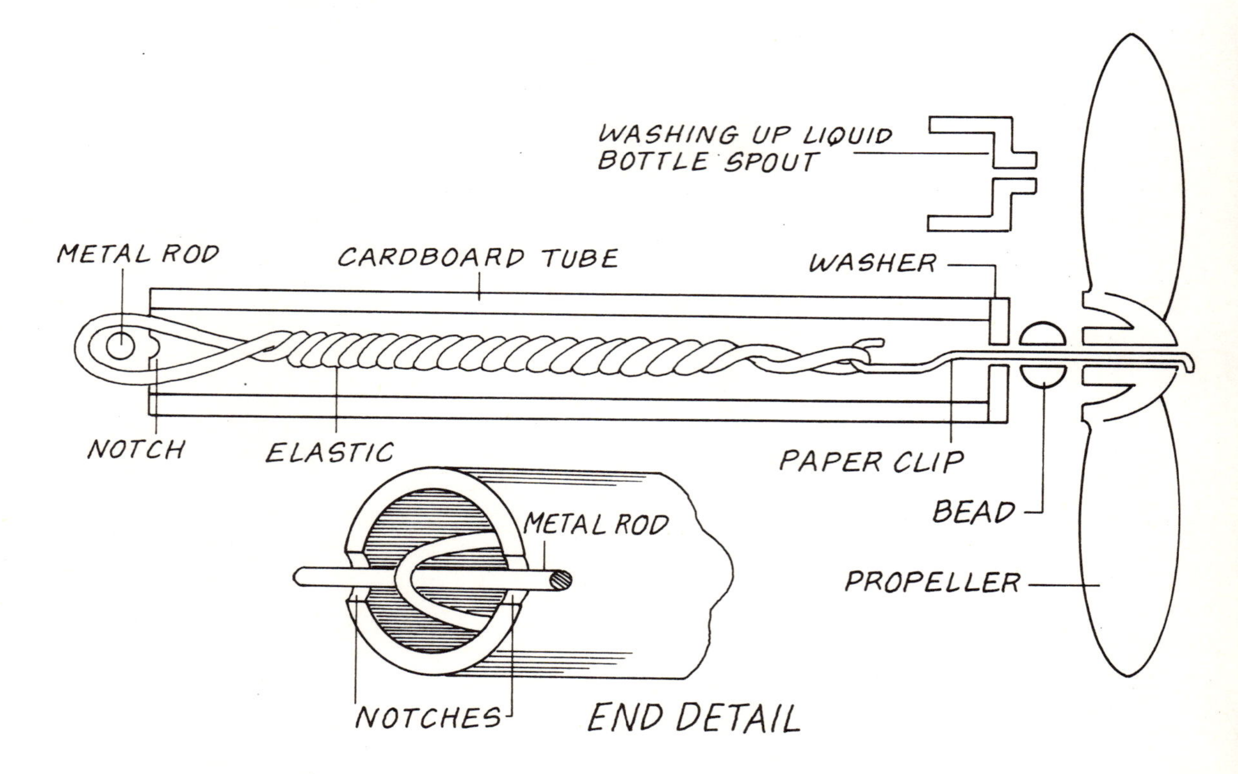

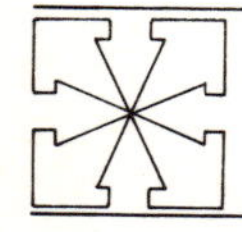

Make a list of all the information sheets you have used. List any other sources used.

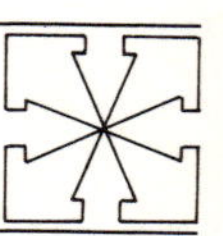

BEC 2 HARD-SHIPS

DESIGN BRIEF 1 Make the rubber-powered boat shown below. Using the test tank, experiment with the angle of bend on your propeller blades until your boat travels as fast as possible in a straight line.

2 Design and make your own boat. It should be no longer than 150 mm and no wider than 100 mm. Your boat should be rubber-powered.

POINTS TO CONSIDER Your boat should be well made, well finished and make good use of colour.

THE TEST

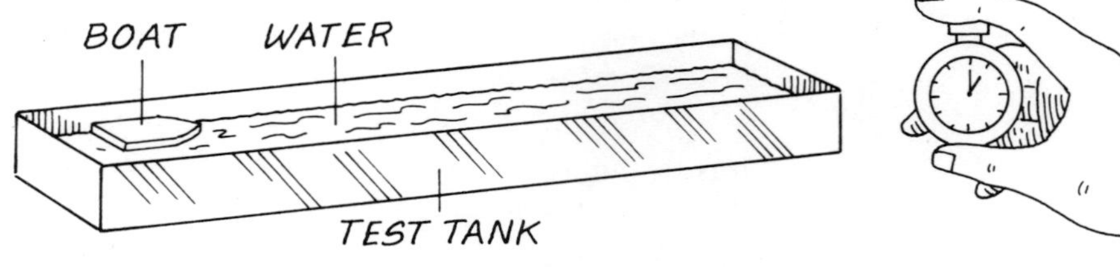

Measure five goes and find the average time your boat takes to travel the length of the tank.

THE TEST BOAT

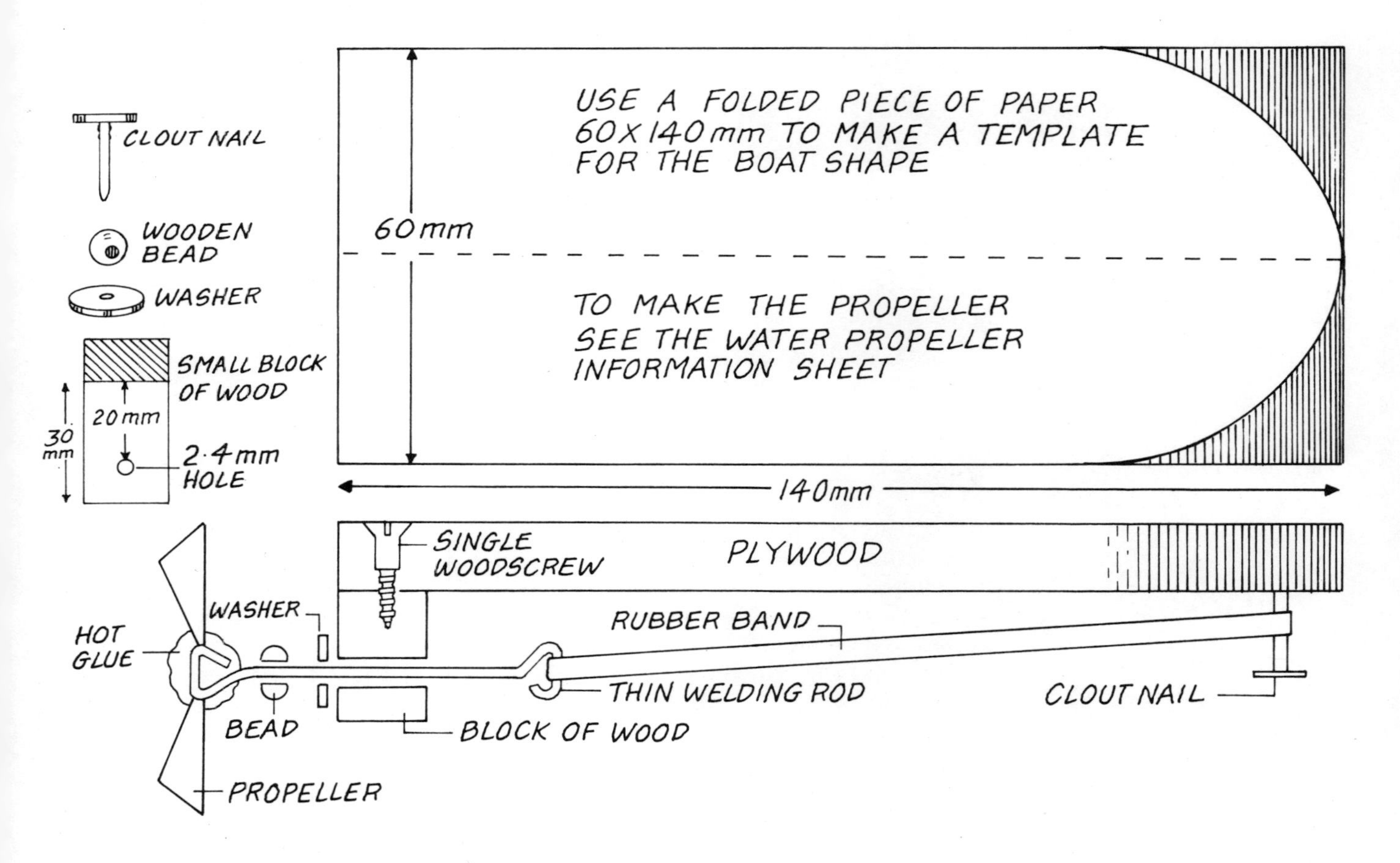

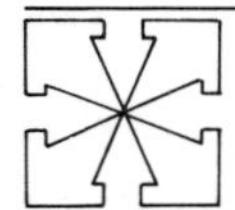

Make a list of all the information sheets you have used. List any other sources used.

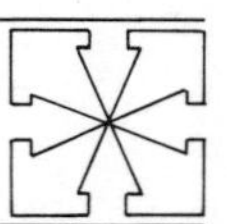

BEC 3 THE RINGCAN SPECIAL

DESIGN BRIEF

Make the Ringcan Special chassis as shown below. When your chassis is running properly, design, make and fit the parts needed to power the chassis. You may use any of the following sources:

1 gravity (e.g. a falling weight);
2 an elastic band;
3 battery power (this must have some form of control, e.g. forward–stop–reverse).

What tests can you think of which will help you to improve the performance of your vehicle?

Now try the Ringcan Special II Brief.

CONSTRUCTION

Chassis pieces

From the wood provided, cut the two chassis side pieces to length (200 mm). Tape them both to the pre-drilled template as shown in the diagram. Drill the holes right through both pieces. Cut the 12 mm diameter dowel to length (see plan sheet). The chassis should now push-fit together (do *not* glue). Cut the axles from 2.4 mm diameter welding rod so they are 30 or 40 mm longer than the width of the chassis.

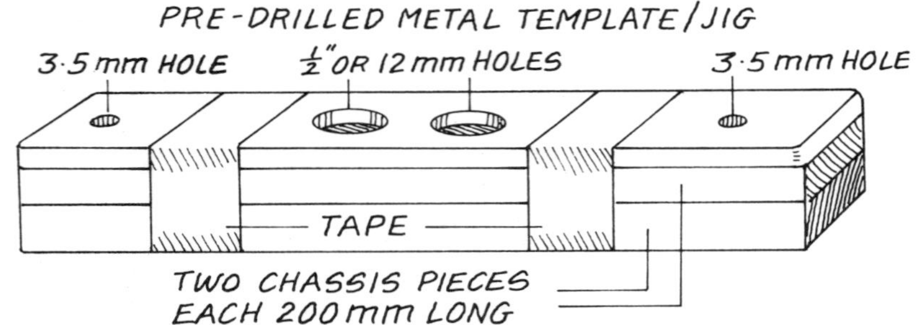

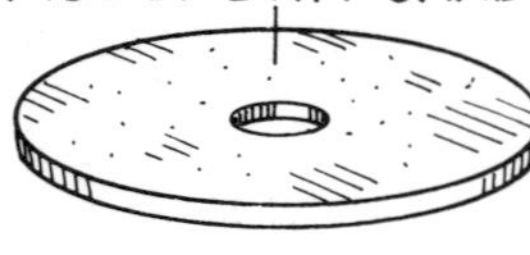

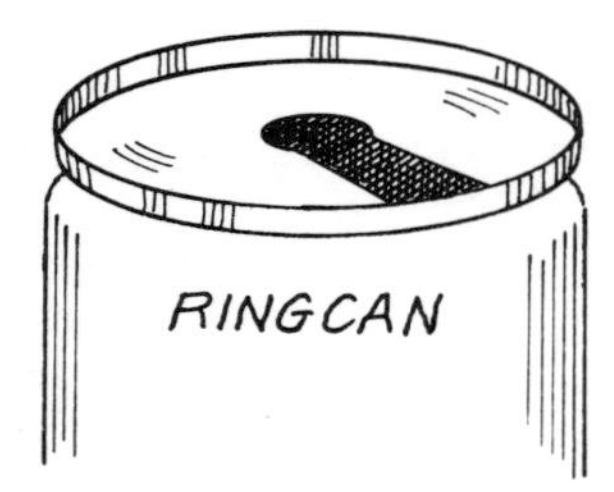

Wheels

Using compasses, mark out a circle radius 30 mm on the card. With a hole punch make a hole exactly in the centre of the disc. Hot glue the disc onto the ring-pull end of the can. Ask your teacher to help you make the hole at the other end of the can. Do the same for the other can. Cut four pieces of tubing 20 mm long as spacers. Fit the wheels and spacers onto the axles and into the chassis. Push the plastic sleeving onto the ends of the axles. Now test your Ringcan Special. Make sure it runs freely and in a straight line.

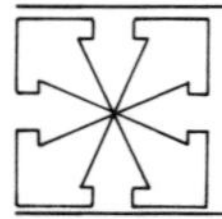

Make a list of all the information sheets you have used. List any other sources used.

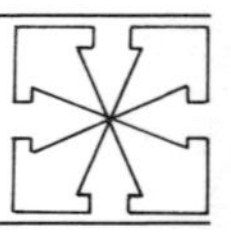

BEC 3*a* THE RINGCAN SPECIAL PLAN SHEET

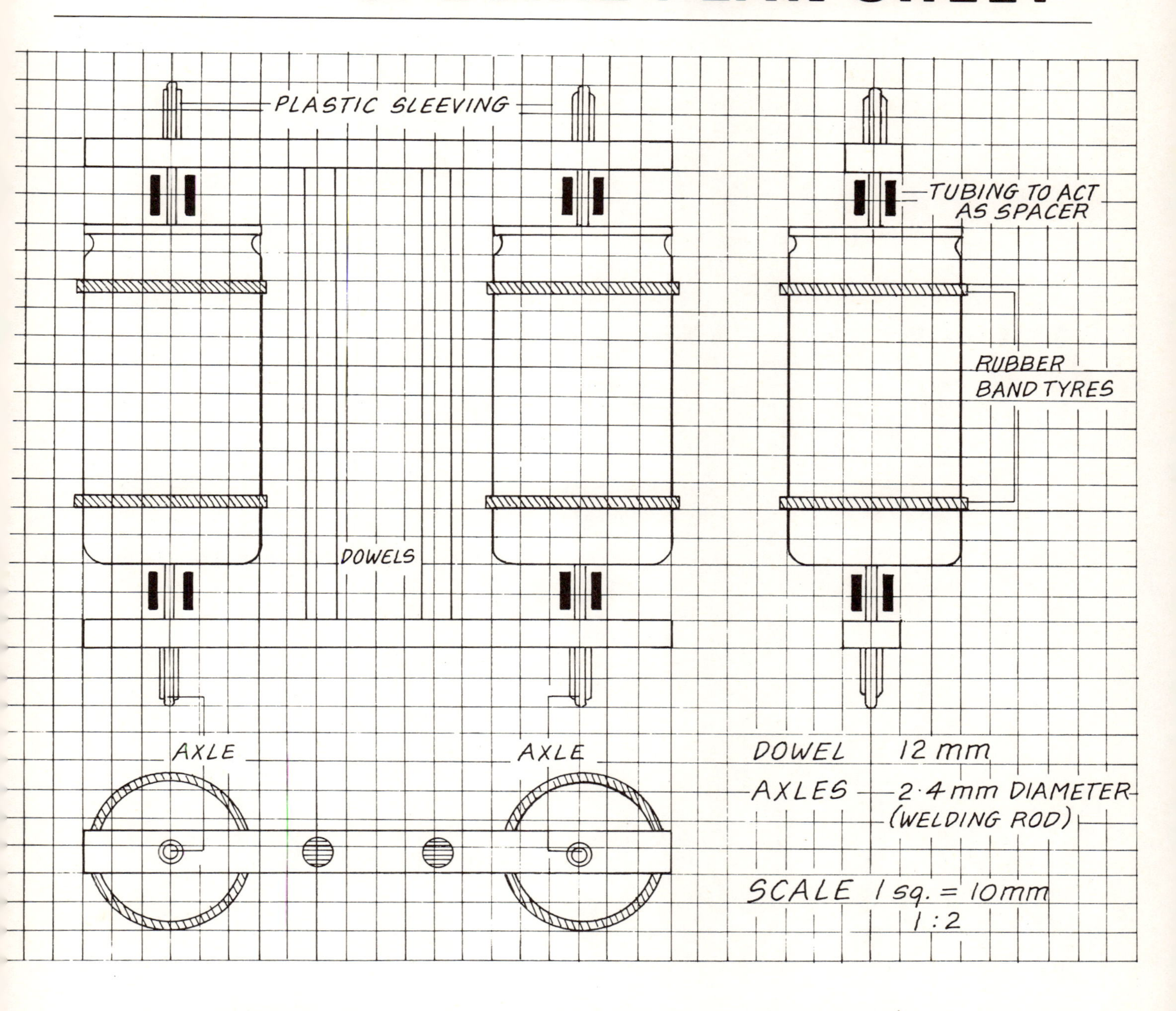

FURTHER INFORMATION

Find two empty ringcans the same size as each other. Measure the height of the cans. Your dowels should be 60 mm longer than the height of the cans,

i.e. for 100 mm cans, the dowels should be 160 mm long,
for 120 mm cans, the dowels should be 180 mm long,
for 150 mm cans, the dowels should be 210 mm long.

NAME

CLASS

BEC 4 RINGCAN SPECIAL II

SUPPLEMENTARY BRIEF

Now you have tried your ideas on the ringcan chassis, design, make and test a vehicle which will do one of the following:

1 climb the steepest slope;
2 go fastest on the flat (how will you stop it?);
3 travel exactly 2 m;
4 travel exactly 2 m in precisely 2 minutes;
5 steer round an obstacle course.

Test and perfect your vehicle (see below). Then design and make an appropriate body shell for your vehicle using corrugated card and papier mâché.

POINTS TO CONSIDER

Your vehicle must be powered by one of the following: battery, rubber band, gravity. If you use a battery then you must have controls (e.g. forward–off–reverse).

Your body shell should be appropriately painted and clear-varnished when dry.

THE TEST

A ramp with an adjustable slope 300 mm wide and 1200 mm long will be provided for climbing. The obstacle course will also be provided by your teacher – ask for details.

You should be able to think of suitable tests for the other types of vehicle, e.g.

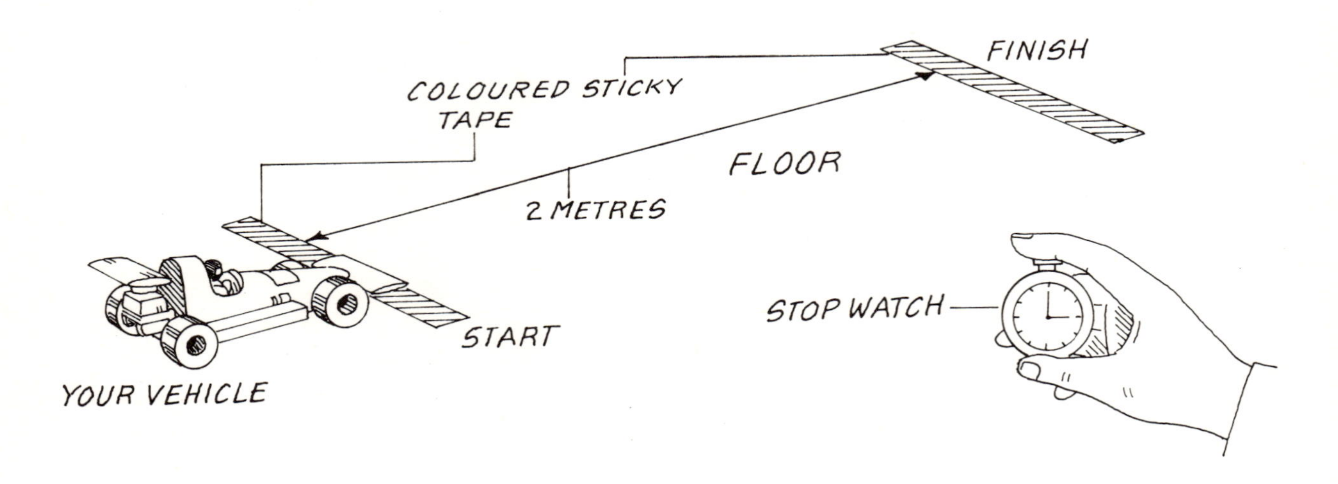

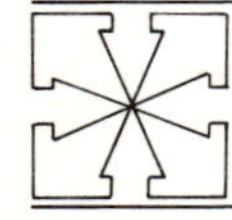

**Make a list of all the information sheets you have used.
List any other sources used.**

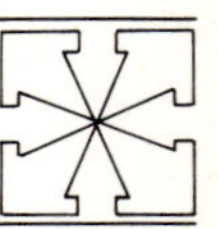

BEC 5 FAIRGROUND FUN

DESIGN BRIEF Design and make a toy based on a fairground machine powered by one of the following:

1 a falling weight;
2 a rubber band;
3 a battery motor.

You may use suitable waste packaging materials, e.g. washing-up liquid bottles, card, cardboard tubes, boxes, margarine tubs, etc.

Your finished design must fit on a baseboard which is 200 × 200 mm and be no more than 300 mm tall.

POINTS TO CONSIDER Find pictures of fairground machines to study the way they are decorated so that you can use the ideas and colours in your own work.

Make a list of all the different fairground machines you can and in each case describe the ways in which they move; e.g. round and round, horizontally, rocking backwards and forwards. Then make a list of things which move in similar ways finding, where possible, how they make that movement; e.g. using a cam.

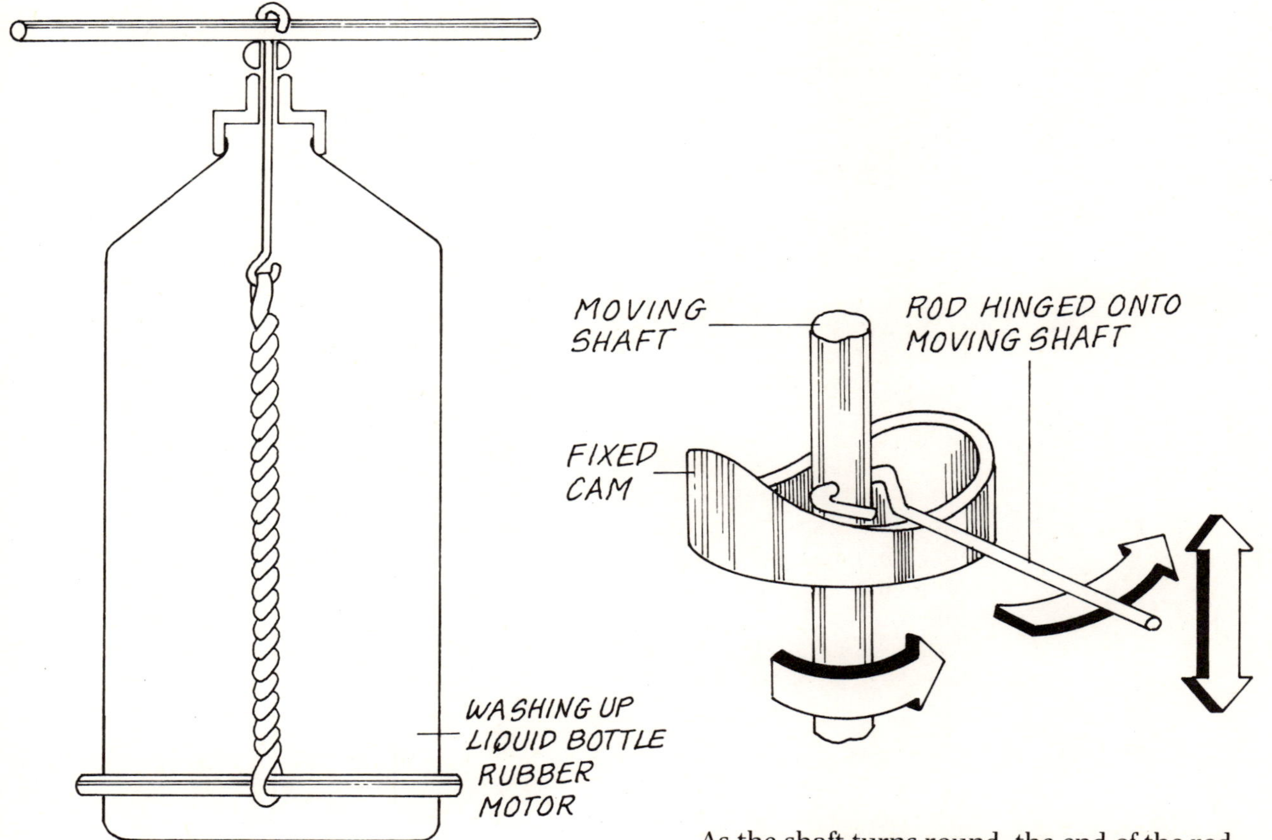

As the shaft turns round, the end of the rod goes up and down as it follows the shape of the cam.

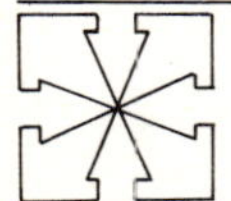

Make a list of all the information sheets you have used.
List any other sources used.

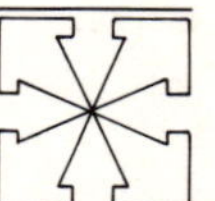

BEL 1 STEADY FREDDY

DESIGN BRIEF

Using the circuit shown below, design and make a 'steady hand' game. You may use only the following materials.

For the case:
a piece of chipboard no larger than 70 × 150 mm;
a piece of acrylic no larger than 200 × 150 mm.

You will also need:
600 mm length of 1.6 mm diameter welding rod;
a short piece of sleeving (for the start and finish);
an LED;
a 680 Ω resistor;
a PP3 battery;
a five-section piece of connector block;
a battery clip;
a length of flexible wire up to 300 mm long.

CIRCUIT DIAGRAM

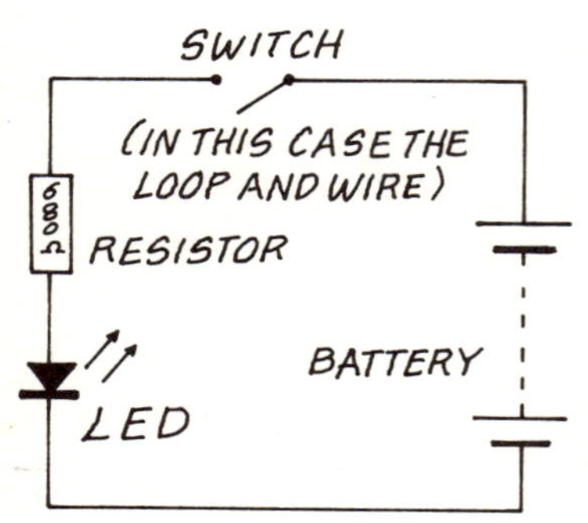

WIRING DIAGRAM

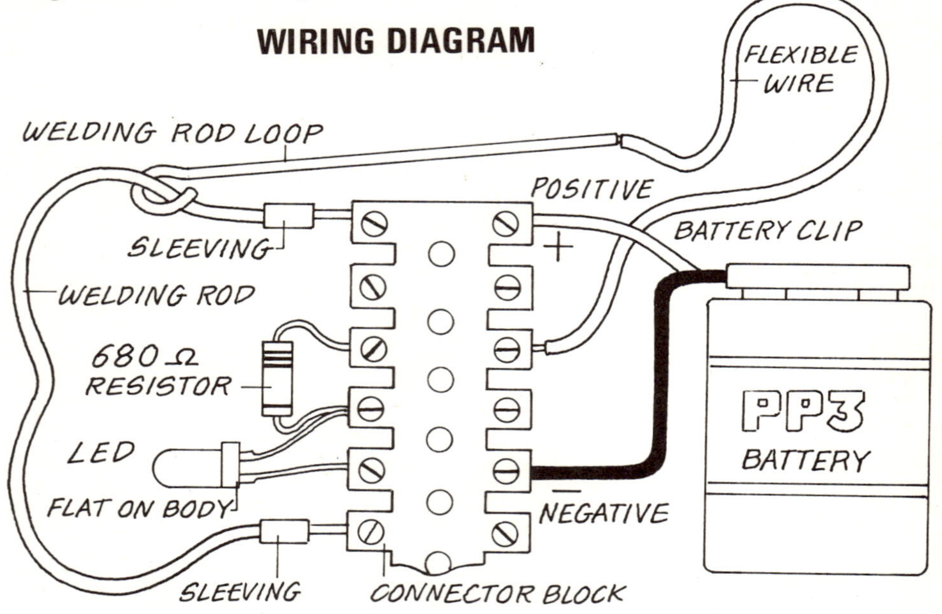

THE TEST

Touch the loop onto the bare part of the welding rod and the LED should light up. If it does not, check that your wiring is correct. Make sure you tighten the connector block screws onto bare wire. The plastic insulation material will not conduct electricity!

Check the LED is the right way round. CAUTION: if you connect the LED directly to battery voltage it will be destroyed!

POINTS TO CONSIDER

The acrylic part of the case should have just two folds.

It should be easy to change the battery.

You could use a 6 V buzzer or a suitable lamp bulb instead of the LED.

A decorated paper sheet on the inside of clear acrylic could make your case more attractive.

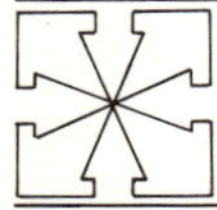

Make a list of all the information sheets you have used.
List any other sources used.

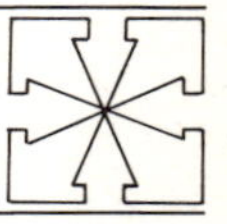

BEL 2 SENSORED

DESIGN BRIEF

Choose one of the sensors shown below and decide upon a suitable application, i.e. what could you use it for? Design a case for the project which is to be made from acrylic. Your case must be no larger than 100 × 60 × 40 mm and should contain the battery. You should be able to change the battery easily.

Sensors

heat wet light
cold dry dark

POINTS TO CONSIDER

The circuit is in three parts. These are the sensor, the trigger circuit and the alarm. Should all three parts be together in the same case, or should some parts be separated from the others? If so, why?

You will need to try out your case design using corrugated card to see if all the controls and connections fit properly and conveniently. This is called a mock-up. You may need to label your controls; how could you do this on acrylic?

Wet and dry sensors you can make yourself

Here is a useful rain sensor made from veroboard (10 track by 24 holes).

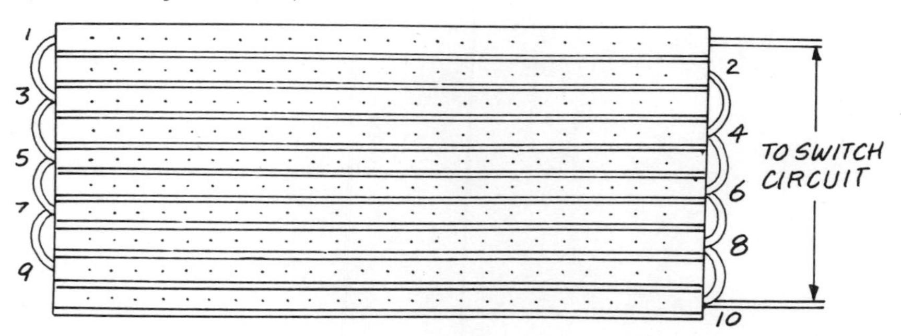

Connect tracks 1 to 3 to 5 to 7 to 9. Connect tracks 2 to 4 to 6 to 8 to 10. Take care! There must be no short-circuits between odd and even tracks.

Here is another useful probe.

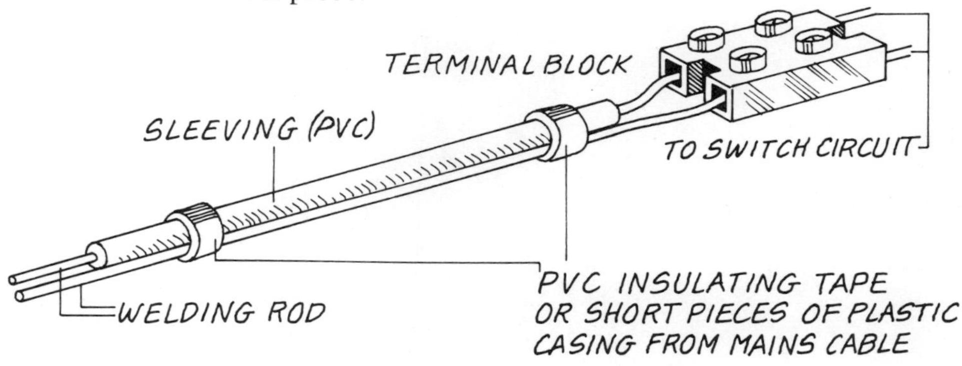

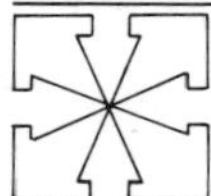

**Make a list of all the information sheets you have used.
List any other sources used.**

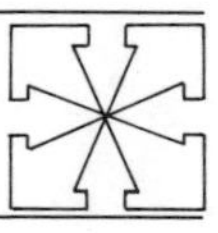

BEL 3 ABOUT TIME

DESIGN BRIEF

Try the timer circuits in the information sheets, then decide upon a suitable application. Design and make a case for your project from acrylic. Your case should be no larger than 100 × 100 × 100 mm and contain the battery. It should be easy to change the battery.

POINTS TO CONSIDER

The best circuit for your timer will depend on the length of the time interval you need. The longer time intervals need a more complex circuit.

Capacitors vary widely from their stated value. This means testing the capacitor in the circuit you are going to use. Check your timer with a stop watch.

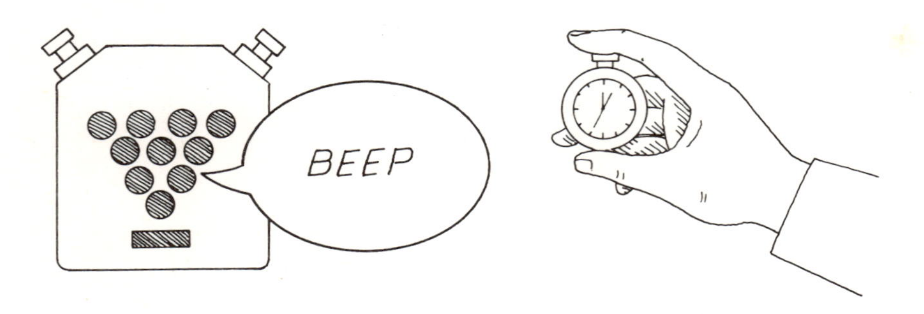

Think carefully about the positioning and labelling of the controls.

If you were marketing your timer, how would you make it attractive to possible customers? Design some suitable advertising material for your timer, e.g. for television or local radio. Design suitable packaging for your product.

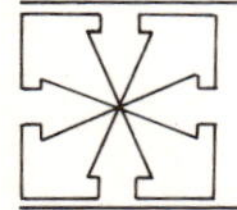

**Make a list of all the information sheets you have used.
List any other sources used.**

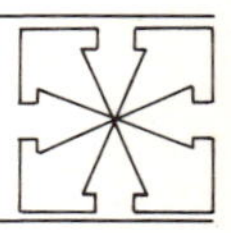

BEL 4 TWEET DEFEAT

DESIGN BRIEF

In many gardens birds do considerable damage to crops. Choosing your materials carefully, design and make a bird-scaring device suitable for the small garden. Your device must only scare and not harm the birds. It must not cause problems for your neighbours.

POINTS TO CONSIDER

How will your bird-scarer frighten – visually or audibly? What scares birds?

Will it be a mechanical device or an electronic one?

How can you protect your device from the damage likely to be caused by the weather?

What source of energy will you use to power your device – wind, solar or battery power?

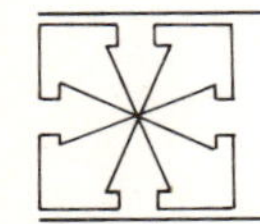

Make a list of all the information sheets you have used. List any other sources used.

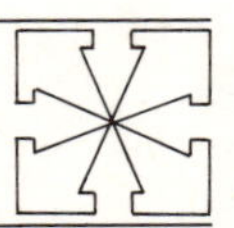

BGP 1 MOBILE MAGIC

DESIGN BRIEF

Babies and young children are often fascinated by brightly coloured, moving objects.

From the materials provided (e.g. 3-ply, hardboard, stiff card or acrylic, plus wire and thread), design and make a mobile for a young person's room. Choose an interesting theme for your mobile: e.g. dog and cats, cat and mice, bees and hive. Make a list of suitable ideas.

POINTS TO CONSIDER

Safety: the pieces must be well finished, i.e. no sharp points or edges; they must be too large to swallow. Wire ends must be made into closed rings.

Your mobile should be colourful, eye-catching and move in an interesting way.

Your mobile should not tangle up!

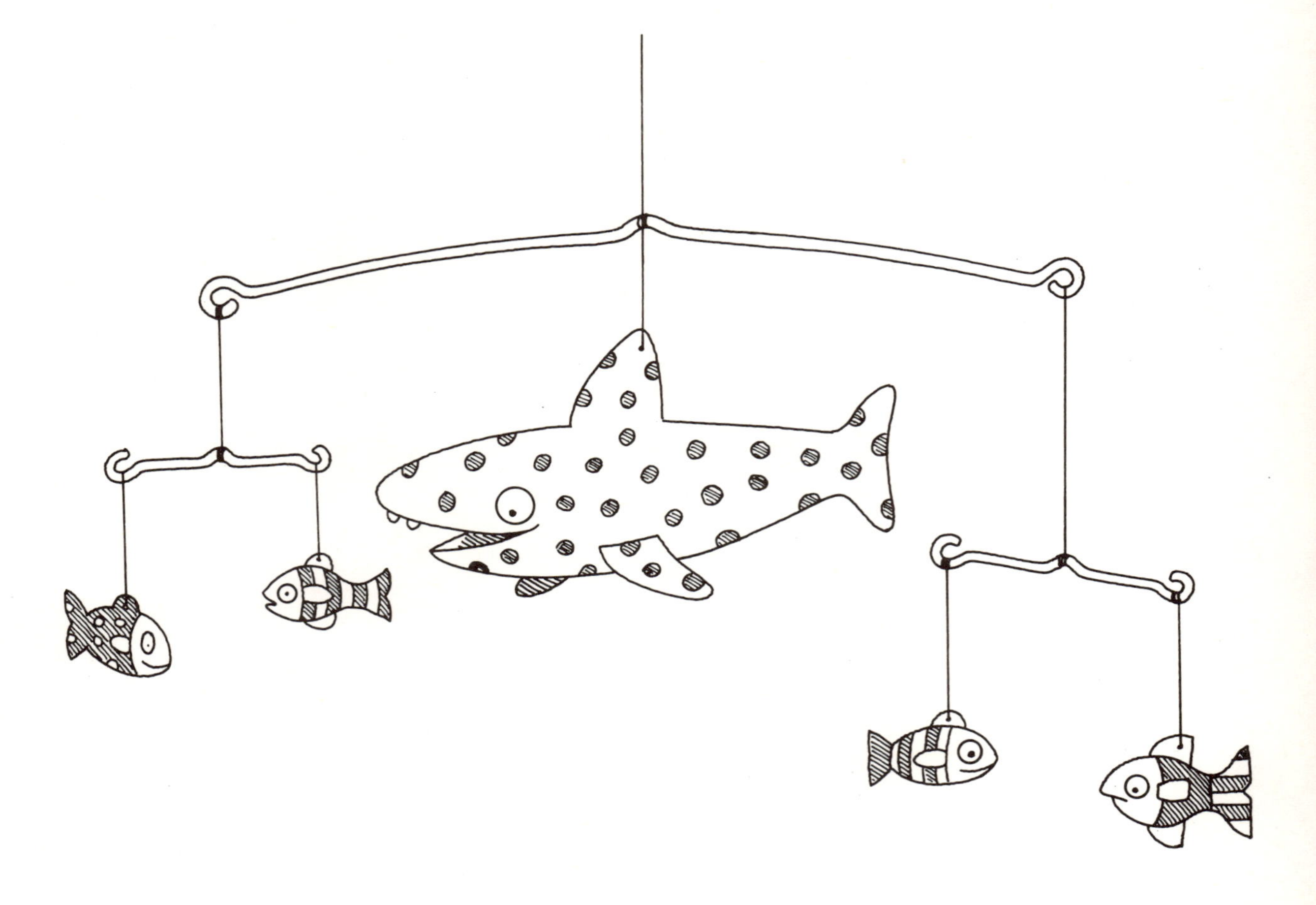

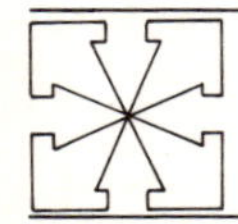

Make a list of all the information sheets you have used.
List any other sources used.

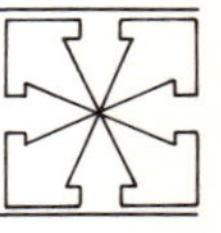

BGP 2 FUN AND GAMES

DESIGN BRIEF Make and decorate the 'spinner' as shown below. Design and make a board game for two or more players. The board, any playing pieces and a container for the pieces must all be made from an A3 (420 × 300 mm) piece of card.

POINTS TO CONSIDER Use paper to make your first attempts (prototypes). Why is this a good idea?

Your finished game should show careful and sensitive use of colour. Try to limit your colour scheme to no more than two colours with black and white. The container for the pieces should be decorated. Try to design a suitable logo.

THE SPINNER Using compasses, radius 2.5 cm, (or a template) draw a hexagon on thin cardboard and number it as shown.

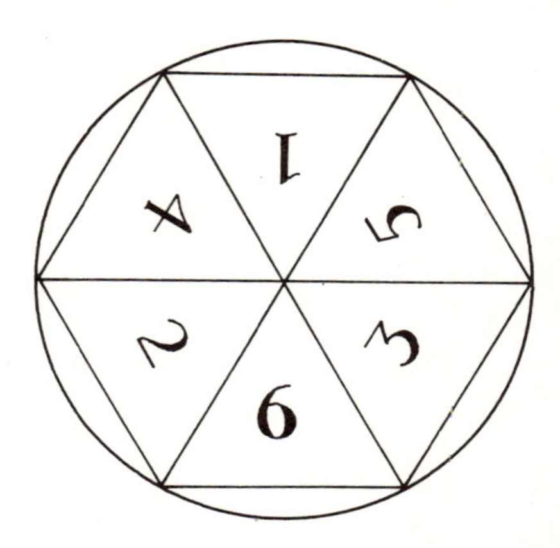

Cut out your hexagon and push half a cocktail stick through the centre. The point should stick out underneath by about 5 mm. Adjust the position of the hexagon until your 'spinner' works smoothly, then fix it with a dab of glue. smoothly, then fix it with a dab of glue.

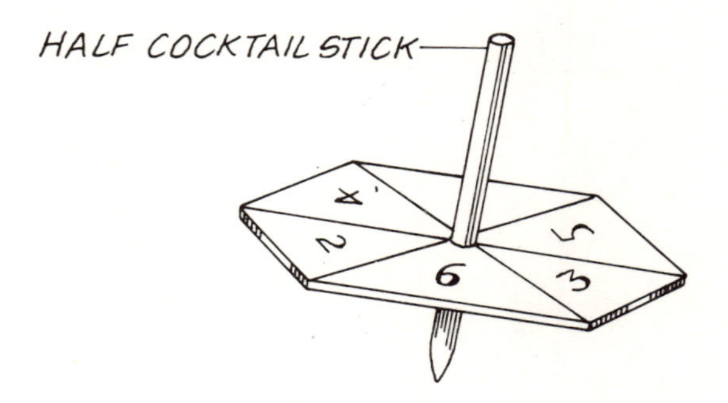

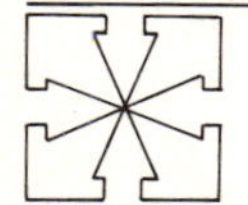

Make a list of all the information sheets you have used.
List any other sources used.

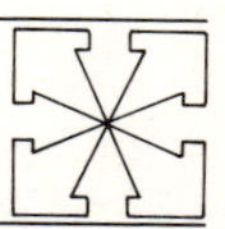

BGP 3 NOUGHTS AND CROSSES

DESIGN BRIEF

From the materials provided, make a lidded box which is 120 mm square.

Design the lid in such a way that the inside or the outside surface can be used as a noughts and crosses board 90 mm square.

Design the playing pieces for the game so they can be made from 25 mm squares of acrylic.

The pieces are to be stored inside the box when not in use.

Use rebate joints to make the corners of your box.

POINTS TO CONSIDER

How many different ways could you make the playing surface?

Do the playing pieces have to be noughts and crosses?

How many of each type of playing piece will you need?

Have you thought of using 'plastic memory' in the design of your playing pieces?

How will you make sure that your game is attractive to look at and pleasant to touch?

Could you design the board and pieces so that a blind person would find it possible to play?

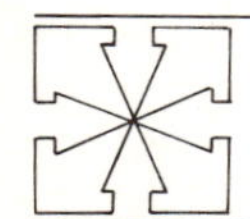

Make a list of all the information sheets you have used.
List any other sources used.

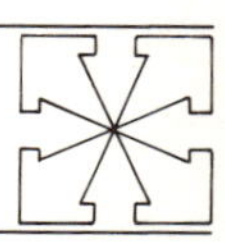

BGP 3a NOUGHTS AND CROSSES PLAN SHEET

This type of plan is called isometric. It is based on 60° angles (the dots on the plan form the corners of equilateral triangles that are arranged in a grid).

Note that in this drawing verticals remain vertical, that horizontal lines are now at 60° to the vertical (or 30° to a true horizontal) and at 120° to each other.

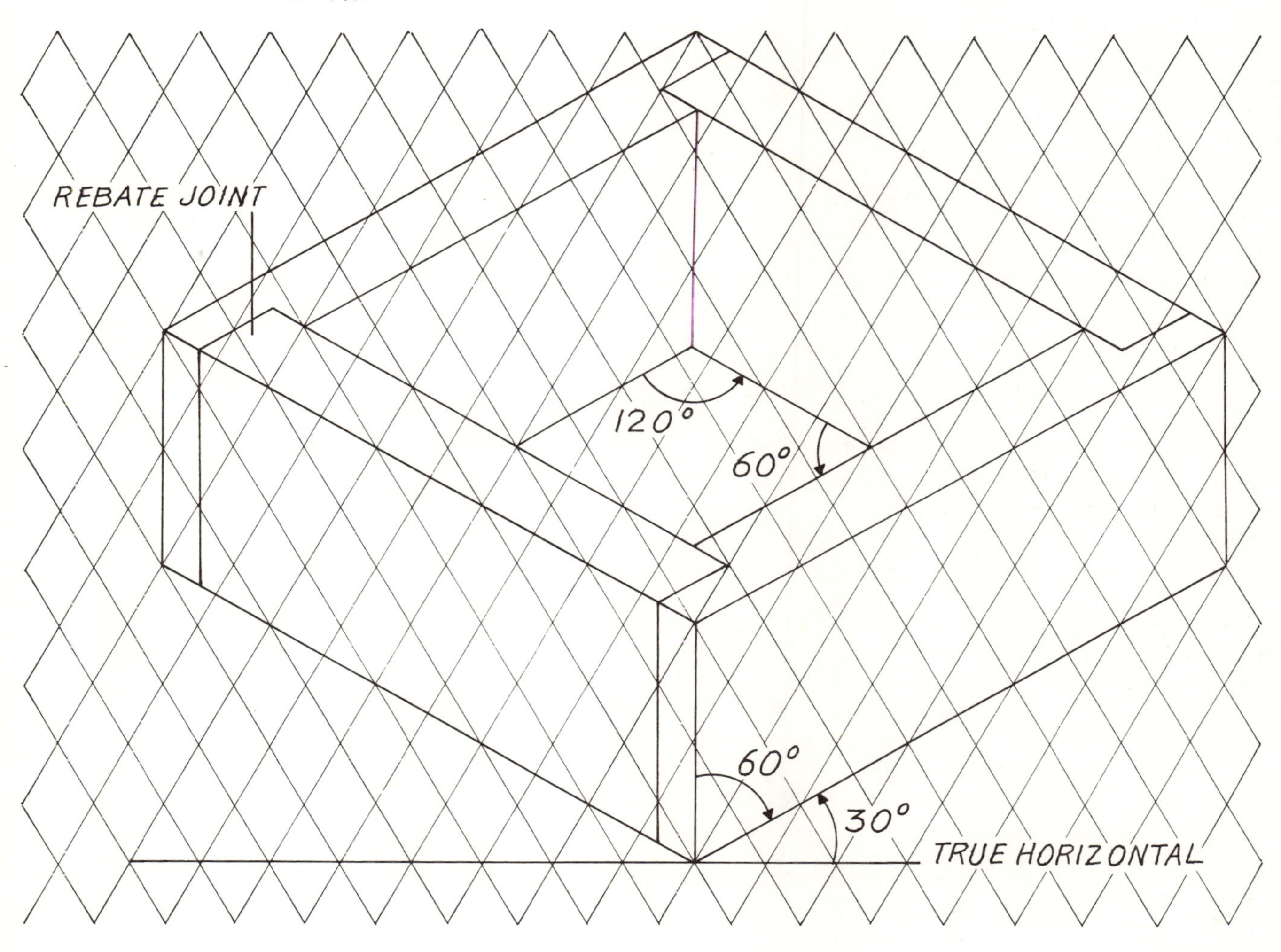

BGP 4 POCKET GAMES

DESIGN BRIEF Using the press mould supplied and no more than four 4 mm diameter ball bearings, design and make an enclosed game using acrylic sheet. The press mould makes a shallow circular dome from clear acrylic 3 mm thick. The space inside the dome is 10 mm deep and 90 mm in diameter.

POINTS TO CONSIDER List as many different types of game as you can. Work out your playing area on a circle of squared paper ruled in 5 mm squares.

Your game should be fun to play – more than once!

It should be pleasant to hold, i.e. no rough edges.

A sensible use of colour will make your game more attractive.

Clear acrylic dome which is glued onto the acrylic disc underneath

This is the playing surface where gates, barriers or holes for ball bearings are put

Paper design coloured in game layout and stuck to the acrylic disc with clear adhesive

Disc of self-adhesive plastic sheet to cover and protect the paper

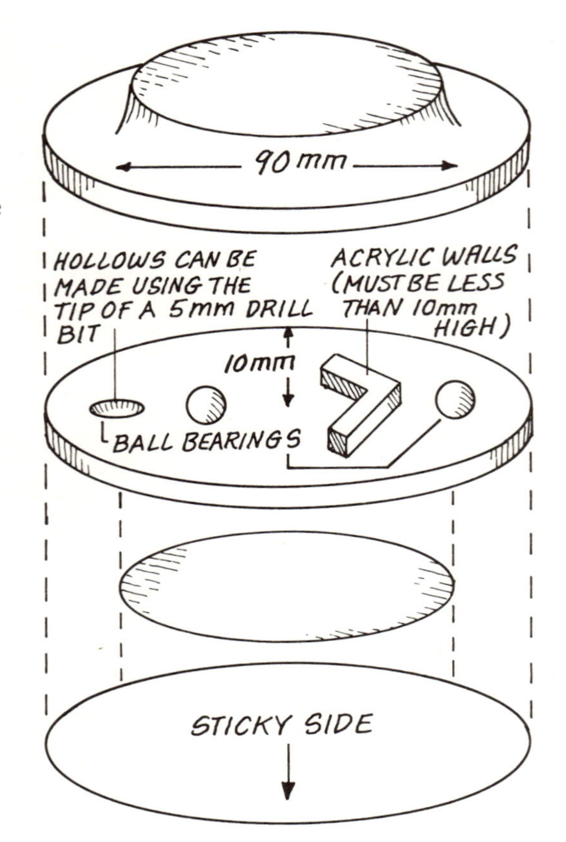

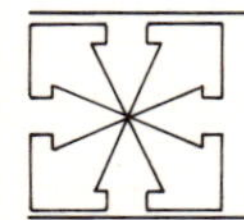

Make a list of all the information sheets you have used.
List any other sources used.

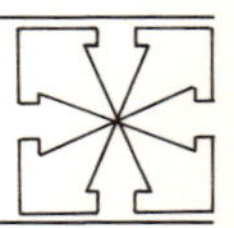

BGP 5 ROCK 'N' ROLL

DESIGN BRIEF

From a piece of plywood 150 × 300 × 10 mm (or chipboard) and a 400 mm length of 2.4 mm welding rod, design and make a toy which uses balance. Here are some examples:

a ballerina
a perching bird
boxers
a gymnast
a somersaulting clown

POINTS TO CONSIDER

Starting with the examples given above, make a list of people or situations showing the use of balance.

From your list choose your best/most original idea.

Your toy should be well finished, colourful and attractive to look at.

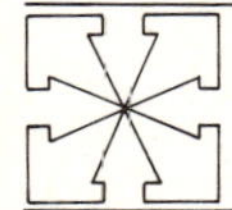

**Make a list of all the information sheets you have used.
List any other sources used.**

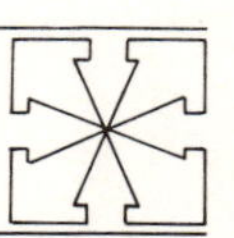

MAKING PIVOTS

INSTRUCTION SHEET 1

MOVING PIVOT

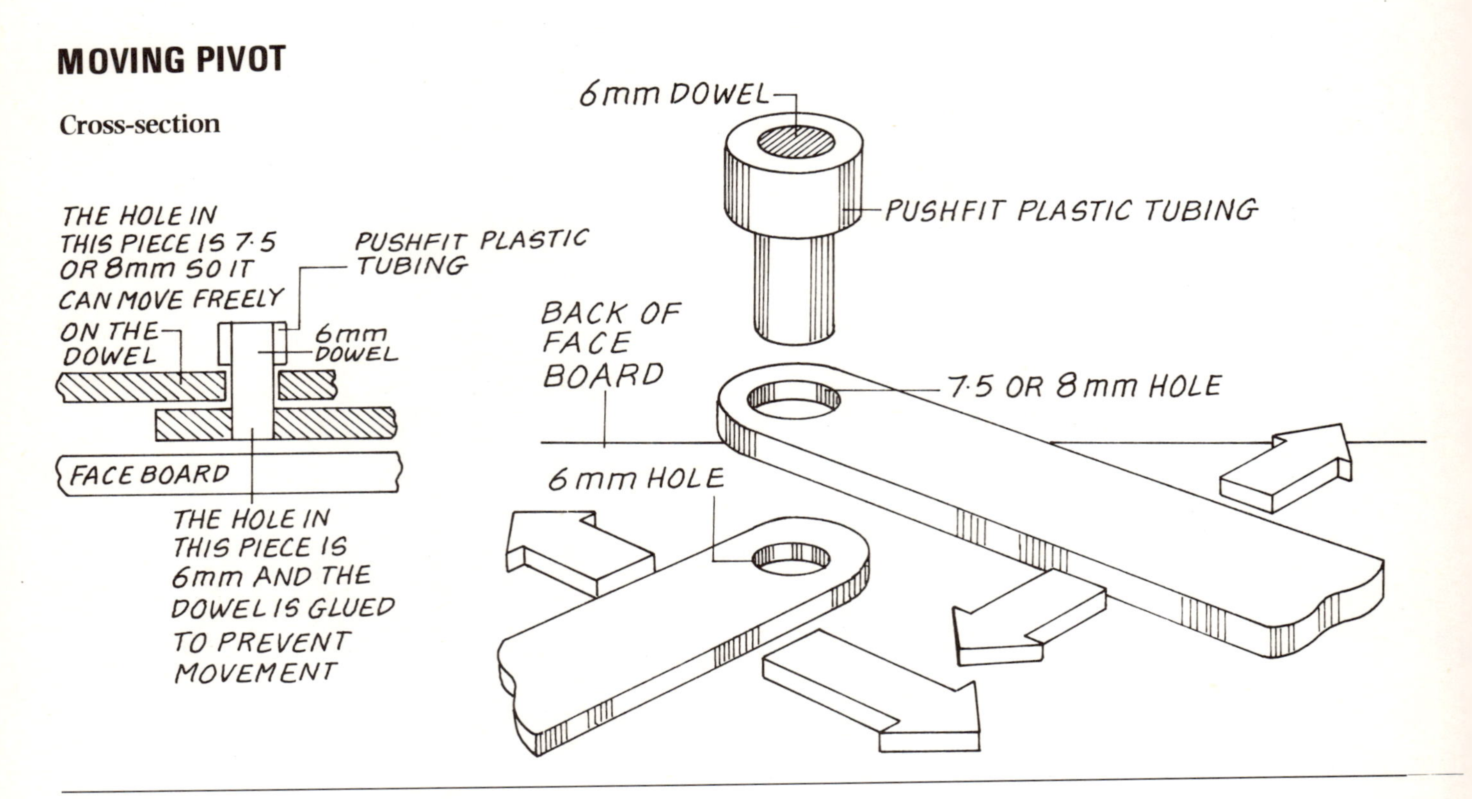

FIXED PIVOT

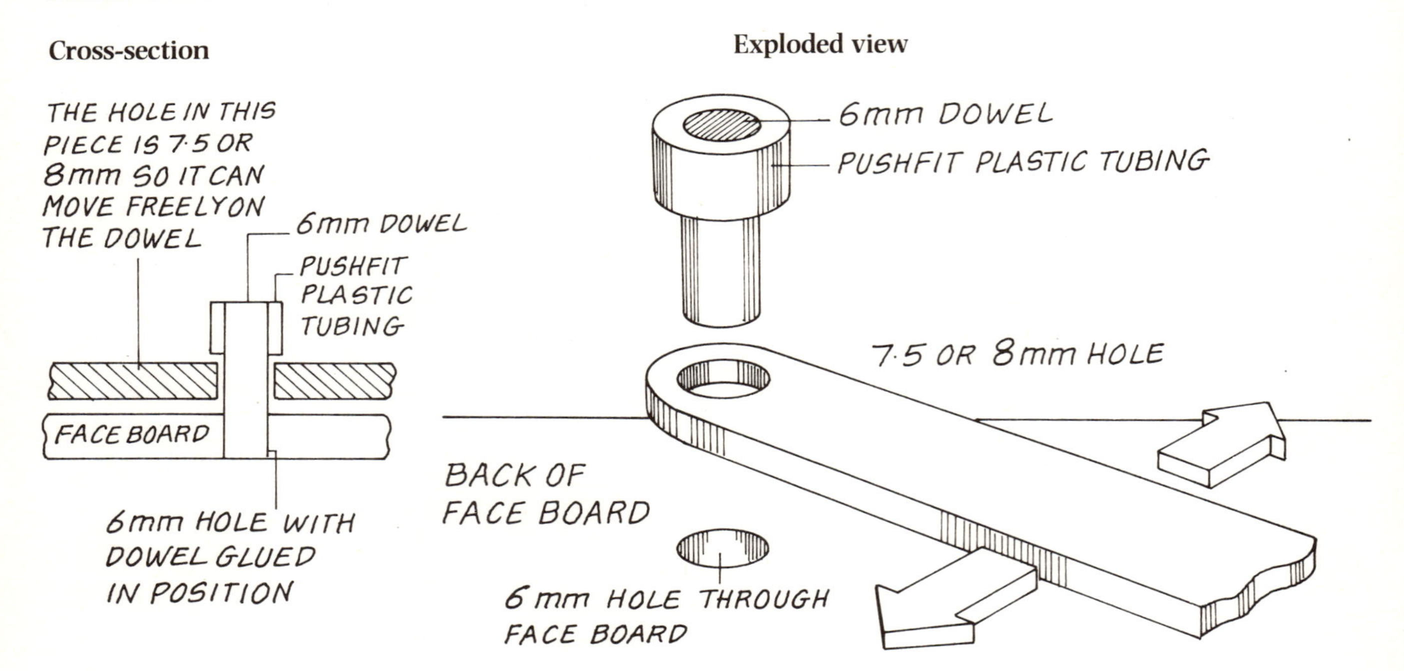

DESIGN BRIEFS

WATER PROPELLER

INSTRUCTION SHEET 2

MAKING THE BLADES

1 Cut a piece of tin plate 40 × 40 mm.

2 Using the template provided, drill two holes as shown.

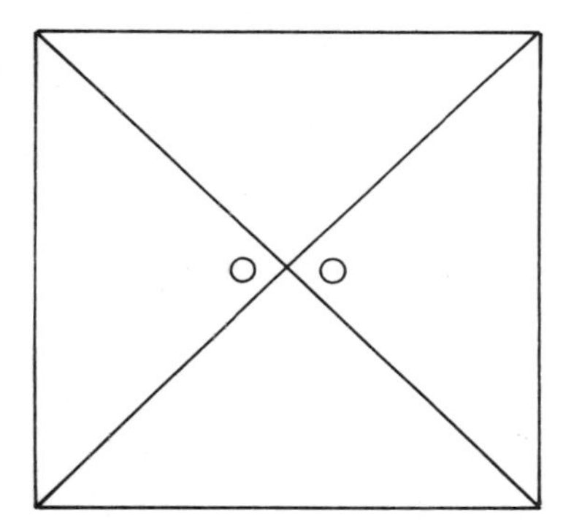

3 With a scriber, mark out the tin plate as shown on the right.

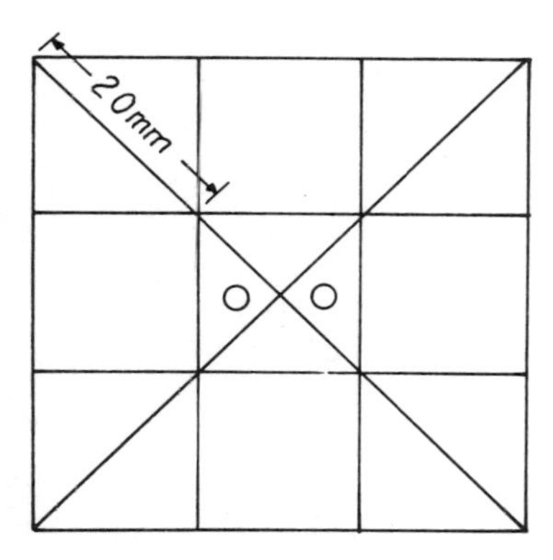

4 Using tin snips, cut out the four triangles as shown.

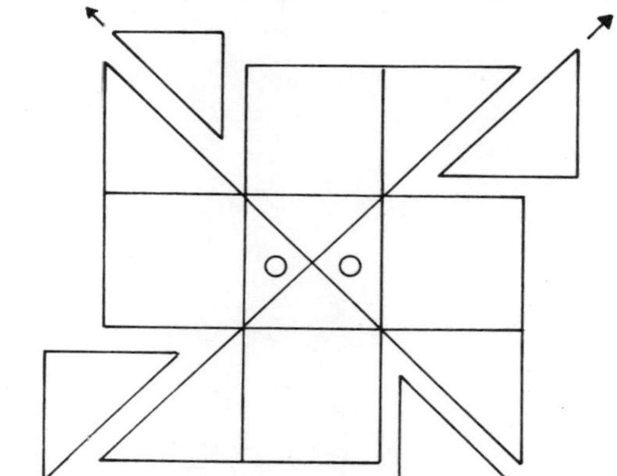

5 Fold along the dotted lines to an angle of 30°.

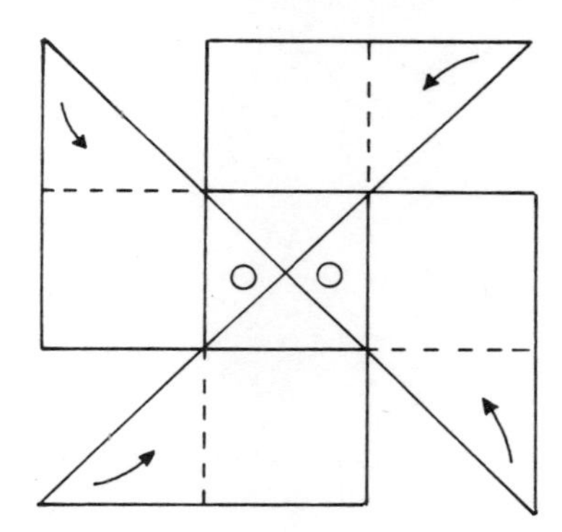

6 Bend a piece of thin welding rod to the shape on the left and insert through the two holes in the tin plate. Now close the end up and straighten the main shaft. Hot glue a blob onto each side of the propeller blade to fix its position.

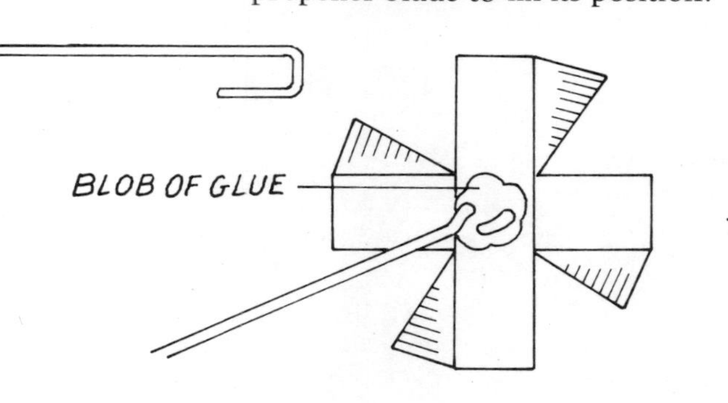

MOUNTING MOTORS

INSTRUCTION SHEET 3

Motors should be mounted carefully so that their drive shafts are, and stay in, the best position to drive the next shaft in the transmission system. This position will depend on the type of transmission used (see Pulleys and Gear Systems) but usually involves the shafts being either parallel or at right-angles.

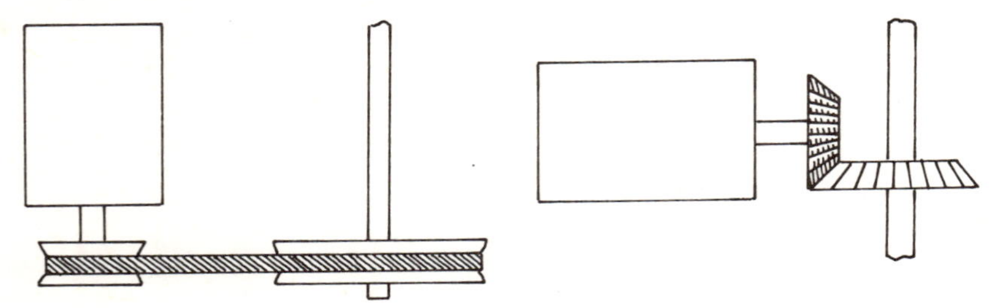

Some motors are made with a mounting plate built in, and so fixing with small screws or bolts is quite simple. Many motors, however, are not fitted with mounting plates and are rather more difficult to use.

METHOD 1

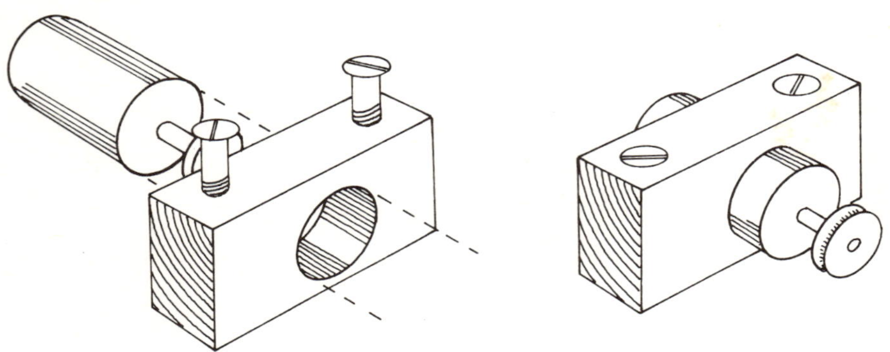

Most motors are cylindrical so they can be mounted in a hole drilled to the correct diameter in a wooden block. (If you cannot drill a hole to fit exactly, then make it slightly too big and wrap tape round your motor till it fits.) You can now fasten the block in position with screws or glue.

METHOD 2

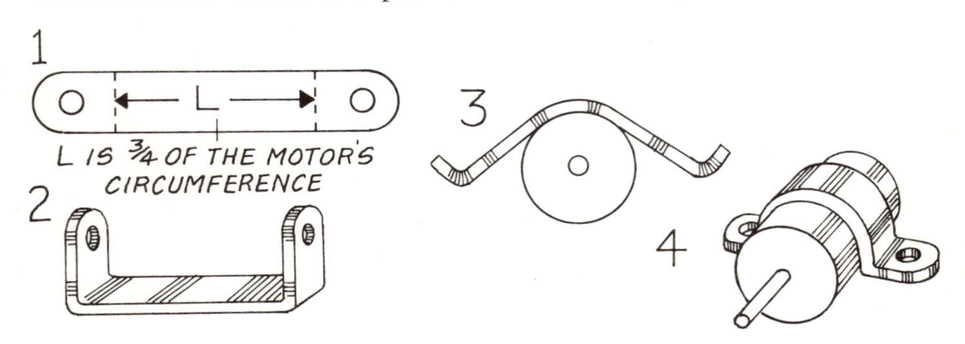

Another way is to make a mounting bracket from thin metal or acrylic. First measure round the circumference of the motor with string or a strip of paper. In figure 1, L is $\frac{3}{4}$ of this circumference. Cut out the shape. Drill the holes before folding up the ends. Mould the bracket around the motor until it is the correct shape. Screw or bolt into position. If your bracket is still loose, wrap tape round the motor till it fits.

BOXES AND LIDS

INSTRUCTION SHEET 4

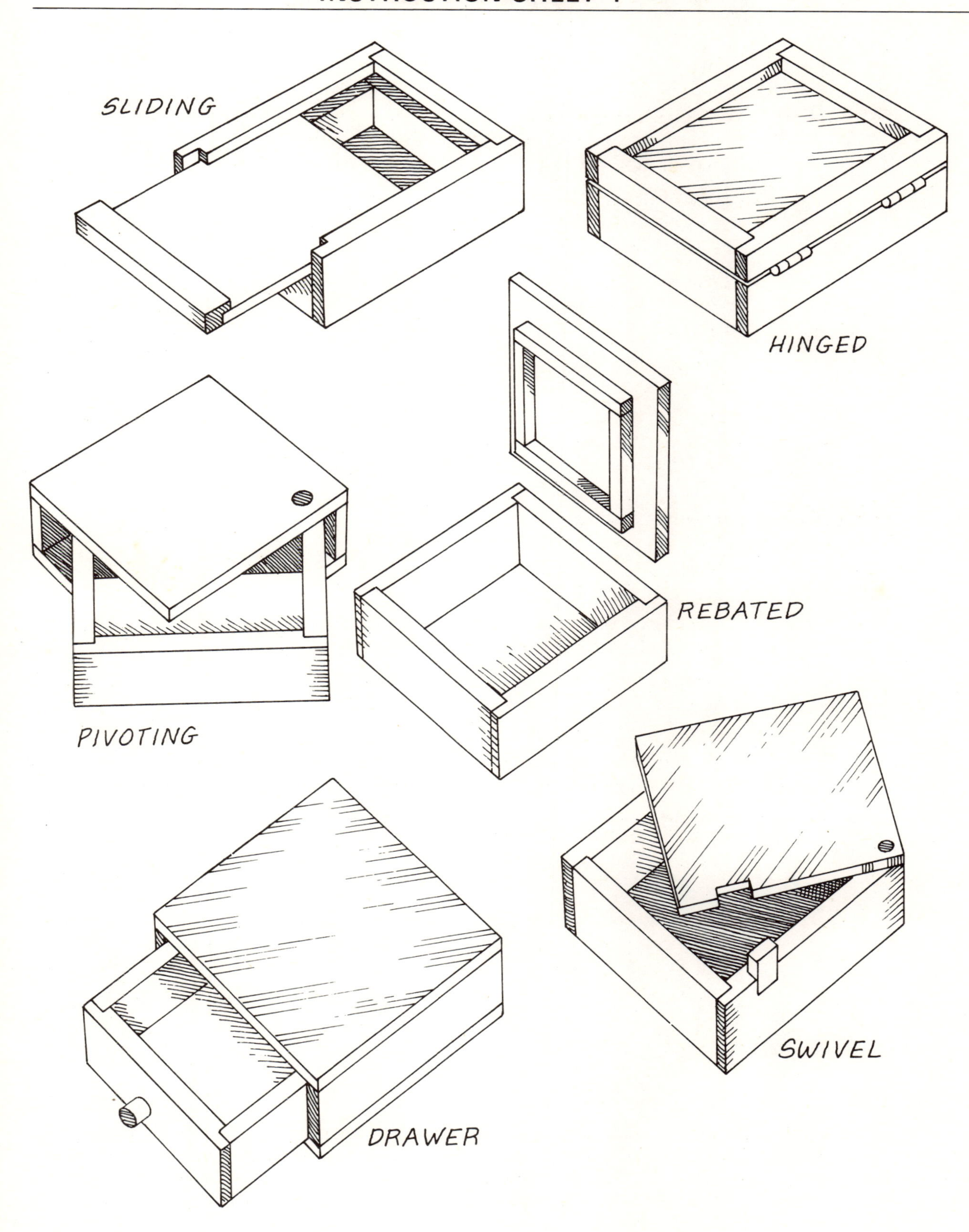

STUDENTS' PROJECT EVALUATION GUIDE

PRACTICAL

1 Fulfilling the Design Brief
To what extent do you think you have succeeded?
In which aspects have you been less successful?

2 Improvements
If you had two more weeks to work on the project, what further improvements would you make?

3 Research
Which areas do you feel you researched well?
In which areas were you short of information?

4 Feedback
What has been the reaction of other people to your work?

5 Skills
Which new skills have you learnt?
Which skills have you improved?

PERSONAL

A Problems
Which were the most difficult parts of the project?
Which were the easiest parts?

B Interest
In which ways did you expect to find the project interesting?
In which ways did you actually find it interesting?

C Learning
What new information, knowledge or skills have you learnt from this project?

D Advice
What advice would you offer to the next group to do this project?
Which changes would you make to this project?

E Observations
Please make any useful observations on points not already covered elsewhere in your project evaluation.

NOTES

Please write your evaluation on the sheet provided. Write the headings and your answers in the same order as above. Please do not forget to put the title of the Design Brief used.

Remember to write your full name and class in the spaces provided.

STUDENTS' PROJECT EVALUATION SHEET

DESIGN BRIEF TITLE

FULL NAME

CLASS

STUDENTS' DESIGN FOLDER GUIDE

THIS SHOULD CONTAIN

1. A list of things you need to find out about before you can work out possible design solutions.
2. Details of any research you have carried out. How and where you have found the information you need.
3. Drawings and notes showing a wide range of possible design solutions.
4. Your reasons for selecting the design solution you have chosen.
5. The plans and drawings for the solution chosen.
6. A planning timetable which shows the order in which you expect you will make each of the parts required. This should give some idea of the time each step will take.
7. A modification or changes sheet where you list any problems which changed the design or the way you intended making the chosen solution. You should give details of what you have changed and why the changes were necessary.
8. When you have completed the project, you should fill in an evaluation sheet. The evaluation sheet should be included in your folder.

SOME HINTS ON PRESENTATION

Do your drawings on separate pieces of paper. Then when they are finished you can try them in different arrangements before you stick them down.

Cut round your work carefully so it makes a frame for each piece. You might 'round off' the corners.

Do your drawings in pencil first and, when they are correct, ink them in with a black drawing pen.

Write your labels in BLOCK CAPITAL LETTERS.

Be selective with your colour scheme. Two or three well-chosen colours are better than a dozen badly chosen ones.

When you put pieces of work on a sheet or in your folder, make sure they are in a sensible order.

Above all, your work should be neat and show clear, well-ordered thinking.